Compressive Force-Path Method

Compressive Force-Path Method

Editor

Sevilin Kartal

Compressive Force-Path Method

Edited by **Sevilin Kartal**

Printed in 2017

ISBN: 978-1-68117-113-5
Library of Congress Control Number: 2015951989

© 2016 by
SCITUS Academics LLC,
616, Corporate Way, Suite 2, 4766,
Valley Cottage, NY 10989

www.scitusacademics.com

Preface

This book presents a method which simplifies and unifies the design of reinforced concrete (RC) structures and is applicable to any structural element under both normal and seismic loading conditions. The proposed method has a sound theoretical basis and is expressed in a unified form applicable to all structural members, as well as their connections. It is applied in practice through the use of simple failure criteria derived from first principles without the need for calibration through the use of experimental data. The method is capable of predicting not only load-carrying capacity but also the locations and modes of failure, as well as safeguarding the structural performance code requirements.

In this book, the concepts underlying the method are presented for the case of simply supported RC beams. The application of the method is progressively extended so as to cover all common structural elements. For each structural element considered, evidence of the validity of the proposed method is presented together with design examples and comparisons with current code specifications. The method has been found to produce design solutions which satisfy the seismic performance requirements of current codes in all cases investigated to date, including structural members such as beams, columns, and walls, beam-to-beam or column-to-column connections, and beam-to-column joints.

Table of Contents

CHAPTER 8   Effects of Sand Quality on Compressive Strength of Concrete: A Case of Nairobi County and Its Environs, Kenya.. 239

CHAPTER 1

Application of the Combined Method for Evaluating the Compressive Strength of Concrete on Site

Samia Hannachi and Mohamed Nacer Guetteche

Civil Engineering Department, Faculty of Engineering Sciences, University Mentouri, Constantine, Algeria

ABSTRACT

Ultrasonic pulse velocity (UPV) and rebound hammer (RH) tests are often used for assessing the quality of concrete and estimation of its compressive strength. Several parameters influence this property of concrete as the type and size of aggregates, cement content, the implementation of concrete, etc. To account for these factors, both of the two tests are combined and their measurements are calibrated with the results of mechanical tests on cylindrical specimens cast on site and on cores taken from the existing structure in work progress at the new-city Massinissa El-Khroub Constantine in Algeria. In this study; the two tests cited above have been used to determine the concrete quality by applying regression analysis models between compressive strength of in situ concrete on existing structure and the nondestructive tests values, the combined method is used, equations are derived using statistical analysis (simple and multiple regression) to estimate compressive strength of concrete on site and the reliability of the technique for prediction of the strength is discussed for this case study.

INTRODUCTION

Evaluation of concrete properties is of great interest, whether to detect altered areasor to control the concrete quality and estimate its compressive strength [1].

The standard methods used to assess the quality of concrete in concrete structures on specimens cannot be considered. The disadvantage is that results are not immediately known, the number of specimens or samples is insufficient for an economic reason, still does not reflect the reality of the structure [2].

The main advantage of nondestructive testing method is to avoid the concrete damage on the performance of building structural components. Additionally, their usage is simple and quick. Test results are available on site. Concrete testing in structures is demanding in which the cores cannot be drilled, where the use of less expensive equipments is required [3].

Several nondestructive evaluation methods have been developed; based on the fact that some physical properties of concrete can be related to the compressive strength of concrete. The Schmidt rebound hammer (SRH) and the ultrasonic pulse velocity (UPV) tests, are combined to develop correlation between hammer/ultrasonic pulse velocity readings and the compressive strength of the concrete. These non-destructive measurements have proved to be an effective tool for inspection of concrete quality concrete.

EVALUATION OF CONCRETE STRUCTURES BY NONDESTRUCTIVE METHODS

The nondestructive testing of concrete has a great technical and useful importance. These techniques have been grown during recent years especially in the case of construction assessment.
All available methods for evaluating in-situ concrete are limited, their reliability is often questioned, and the combination of two or more techniques is emerging as an answer to all these problems [4].

The combination of several techniques of nondestructive testing is often implemented empirically, combining two techniques most often used to enhance the reliability of the estimate compressive strength of concrete; the principle is based on correlations between observed measurements and the desired property [5].

The compressive strength of concrete is usually the most sought after property. This is leads to the development of a method that combines index rebound hammer and the ultrasonic pulse velocity UPV [1].

The objective of the combined tests is to evaluate the compressive strength of concrete in situ; the best approach is generally to develop a relationship of correlation between the upv/the index of rebound hammer and the compressive strength of standardized laboratory specimen, in some cases specimen are not available, then a number of cores must be taken to establish this relationship [6]. The standardized combined method the most widely used is SonReb method developed by RILEM [7], first born and established in Romania then developed in Australia and in Europe. The improvement in the reliability of the measures is explained by taking into account the contradictory effect of variability factors of some properties for each of the two techniques (ultrasonic pulse velocity/rebound hammer).

Rebound Hammer (RH) Test EN12504-2
The rebound (Schmidt) hammer is one of the oldest and best known methods. It is usually used in comparing the concrete in various parts of a structure and indirectly assessing concrete strength. The hammer weighs about 1.8 kg and is suitable for use both in a laboratory and in the field. The rebound of an elastic mass depends on the hardness of the surface against which its mass strikes.

The test is described in ASTM C805 and EN12504- 2:2001. The results of rebound hammer are significantly influenced by several factors [4,6] such as: smoothness of test surface; size, shape, and rigidity of the specimens; age of the specimen; surface and internal moisture conditions of the concrete; type of coarse aggregate; type of cement; carbonation of concrete surface.

According to EN13791:2003 standard rebound hammer test with calibration by means of cores test may be used for assessment of in situ concrete strength. In situ strength can be estimated using a basic relationship with a determined factor for shifting the basic relationship curve to take into account of the specific concrete and production procedure [6].

Ultrasonic Pulse Velocity (UPV) EN 12504-4(a) Rebound Hardness Test

The method consists of measuring the ultrasonic pulse velocity through the concrete with a generator and a receiver. The tests can be performed on samples in the laboratory or on-site. Many factors affect the results, the surface and the maturity of concrete, the travel distance of the wave, the presence of reinforcement, mixture proportion, aggregate type and size, age of concrete, moisture content, etc., furthermore some factors significantly affecting UPV might have little influence on concrete strength [8].

The test is described in ASTM C597, EN12504-4:2004, a critical comparison of several standards from different countries is given in a review paper by Komlos [9].

Core Sampling in Situ EN12504-1

Coring is a direct measure of the in-situ strength of concrete. It is mainly used to provide a calibration of an indirect method and rarely used for determining the rate of strength gain.

Core sampling is a destructive test which is used to evaluate the suspicious concrete. At least 3 core samples should be taken from each area. The height of core cylinder is 2.0 diameters and maximum size of aggregate is 1/4 diameter or less.

Due to unknown effects of reinforcing bars in the samples and also in order to keep the integrity of structure, it is better to provide bar detection process before coring and to remove the bars from samples before putting them in compression machine. Any visual defect of concrete should be recorded before compression test and should be applied in analysis [10].

EXPERIMENTAL METHODOLOGY

In this study data from destructive testing and nondestructive testing are provided by the laboratory—CTC-Constantine (Algeria body control) for monitoring and quality control of structures in work progress.

The objective of this work is to study the reliability of these nondestructive techniques and identify factors that affect the interpretation of their results.

The interpretation of results is given through a combination of correlations between nondestructive techniques and those of mechanical tests.

Correlations between compressive strength at 28-day specimens (16/32) cast on site for a concrete proportioned to 350 kg/m³ and stored in air and measures of non-destructive testing in case the hammer and ultrasonic Correlations between compressive strength of cores from the concrete structure (Cores with a diameter of 6.5 cm and a height of 13 cm.

All data from both destructive tests and nondestructive test are resumed in Tables 1 and 2.

COMBINED METHOD

The reduction of the influence of several factors affecting rebound hammer test and UPV method could be partially achieved by using both methods together. A classical example of this application is the SONREB method, developed mostly by the effort of RILEM Technical Committees 7 NDT and TC-43 CND [7] and widely adopted in Romania [11]. The relationship between UPV, rebound hammer and concrete compressive strength are there given in the form of a monogram.

The improvement of the accuracy of the strength prediction according to Facaoaru [11] is achieved by the use of correction factors taking into account the influence of cement type, cement content, petrologic aggregate type, fine aggregate fraction, and aggregate maximum size.

Table 1. Test data for specimens.

Level	Element	Rebound value N	V (upv) m/s	28 days compressive strength (MPa)
GF	Pile	29	3980	34.5
RDC	Pile	30	4100	34.0
RDC	Pile	26	3870	34.0
RDC	Pile	28	3950	36.0
RDC	Beam	30	4180	36.0
RDC	Beam	30	4090	33.0
1stfloor	Pile	22	3730	37.0
1stfloor	Pile	27	3890	35.5
1stfloor	Pile	26	3850	34.0
2^{d}floor	Pile	23	3780	38.5
2^{d}floor	Pile	26	3810	37.5
2^{d}floor	Pile	26	3800	36.5
3^{d}floor	Pile	22	3720	35.0
3^{d}floor	Pile	22	3710	37.0
3^{d}floor	Pile	23	3780	36.0
3^{d}floor	Beam	30	4130	25.5
3^{d}floor	Beam	30	4190	27.0
3^{d}floor	Beam	30	4160	26.5

Table 2. Test data for cores.

Level	Element	Rebound value N	V (upv) m/s	28 days compressive strength (MPa)
GF	Pile	26	3860	13.0
RDC	Pile	25	3830	12.5
RDC	Pile	27	3890	14.6
RDC	Pile	28	3920	15.8
RDC	Beam	24	3790	12.7
RDC	Beam	26	3870	13.5
1stfloor	Pile	27	3890	9.2
1stfloor	Pile	22	3730	9.6
1stfloor	Pile	26	3850	14.7
2^{d}floor	Pile	24	3780	16.5
2^{d}floor	Pile	25	3840	17.6
2^{d}floor	Pile	26	3890	13.0
3^{d}floor	Pile	23	3780	12.5
3^{d}floor	Pile	25	3710	12.4
3^{d}floor	Pile	24	3720	11.8
3^{d}floor	Beam	27	3900	13.0
3^{d}floor	Beam	28	3930	15.2
3^{d}floor	Beam	28	3940	13.7

The accuracy of the combination of rebound hammer and ultrasonic pulse velocity results in improved accuracy in estimating the compressive strength of concrete.

The interest in using the combined technique is that the variability of certain properties of concrete produce opposite effects for each of its components (hammer and UPV). For example a raise in moisture increases the value of the ultrasonic pulse velocity but decreases the value of the rebound hammer.

Several linear and nonlinear multiple correlation equations have been developed (**Table 3**) and are available in literature: Tanigawaand et al.

1984; Malothra and Carino, 1991; Qasrawi, 2000; Ariogluand et al. 2001 [12].

STATISTICAL ANALYSIS

All research has shown that petrographic property of aggregates and the composition and means of implementation of concrete significantly influences the values of the rebound hammer and the ultrasonic pulse velocity. To overcome erroneous results and improve the reliability of estimating the compressive strength by nondestructive testing, it is essential to calibrate the measurements from the two trials.

The statistical analysis seems to be the solution to interpret the observed data using the results of mechanical tests on specimens and cores and is needed to calibrate these tests.

In this study, relationships between the results of mechanical testing of specimens and cores and those from non-destructive testing (UPV and RH) are established, the values are plot in graphs, and Matlab is used to extract the curves (regression line), and R^2: determination coefficients are obtained for each regression line.

Simple linear regression is adopted to obtain the correlations below: Figures 1-6.

Resistance RES (dependent variable)-vs-N; hammer index (independent variable).

Resistance RES (dependent variable)-vs-V; ultrasonic velocity (independent variable).

Multiple linear regressions is adopted to find correlations Resistance RES (dependent variable)-vs-Index hammer and ultrasonic velocity (N, V) (independent variables).

In parallel with the statistical analysis; other procedures are needed to analyze the results of correlations; standards and specifications are developed [4,13] an none has achieved consensus as hoped.

Tables 4 and 5 summarize regression equations and the values of determination coefficients for both the correlations made for cylindrical specimen and cores.

CONCLUSIONS AND DISCUSSION

Based on this study, it appears that using more than one non-destructive technique provides a better correlation and in this sense contributes to more reliable strength evaluation of concrete.

Table 3. Equations of existing relationship used for compressive strength estimation of concrete [12].

Eq. No.	Equations	Explanations	Reference	RMSE
	Single-variable equations			
1	$fc = 21.575 \times L - 72.276$	fc[MPa], L[cm]	NDT Windsor Sys. Inc.(1994)	3.7813
2	$fc = 1.2 \times 10^{-5} \times V^{1.7447}$	fc[MPa], V[km/s]	Kheder 1 (1998)	6.0974
3	$fc = 0.4030 \times R^{1.2083}$	fc[MPa]	Kheder 2 (1998)	2.1651
4	$fc = 36.72 \times V - 129.077$	fc[MPa], V[km/s]	Quasrawi 1 (2000)	3.6981
5	$fc = 1.353 \times R - 17.393$	fc[MPa]	Quasrawi 1 (2000)	2.8152
6	$fc = -5333 + 5385 \times L$	fc[MPa], L[in]	Malhotra *et al*.	2.2128
	Multi-variable equations			
7	$c = -25.568 + 0.000635 \times R^3 + 8.397V$	fc[MPa], V[km/s]	Bellander (1979)	2.2128
8	$fc = -24.668 + 1.427 \times R + 0.0294V^4$	fc[MPa], V[km/s]	Meynink *et al*. (1979)	7.0654
9	$fc = 0.745 \times R + 0.951 \times V - 0.544$	fc[MPa], V[m/s]	Tanigawa *et al*.	2.1000
10	$fc = [R/(18.6 + 0.019 \times R + 0.515 \times V)]$	fc[kg/cm^2], V[km/s]	Postacioglu (1985)	3.7617
11	$11 fc = 18.6 \times e^{0.019 \times R - 0.513V}$	fc[kg/cm^2], V[km/s]	Arioglu *et al*. (1991)	2.9205
12	$fc = 10^{3.110} \sqrt{logR^3 \times V^4}^{-5.890}$	fc[kg/cm^2], V[km/s]	Arioglu *et al*. (1994)	4.2305
13	$fc = -39.570 + 1.532 \times R + 5.0.614 \times V$	fc[kg/cm^2], V[km/s]	Raynar *et al*. (1996)	7.5910
14	$fc = 0.00153 \times (R^3 \times V^4)^{0.611}$	fc[kg/cm^2], V[km/s]	Arioglu *et al*. (1996)	11.1623
15	$fc = 0.0158 \times V^{0.8254} \times R^{1.1371}$	fc[kg/cm^2], V[km/s]	Kheder 3 (1998)	2.1375

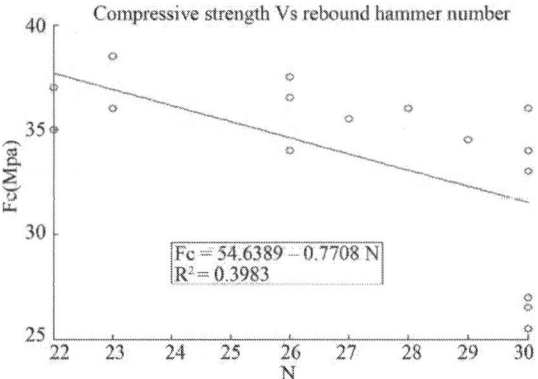

Figure 1. Correlation: compressive strength-rebound number (cylindrical specimens).

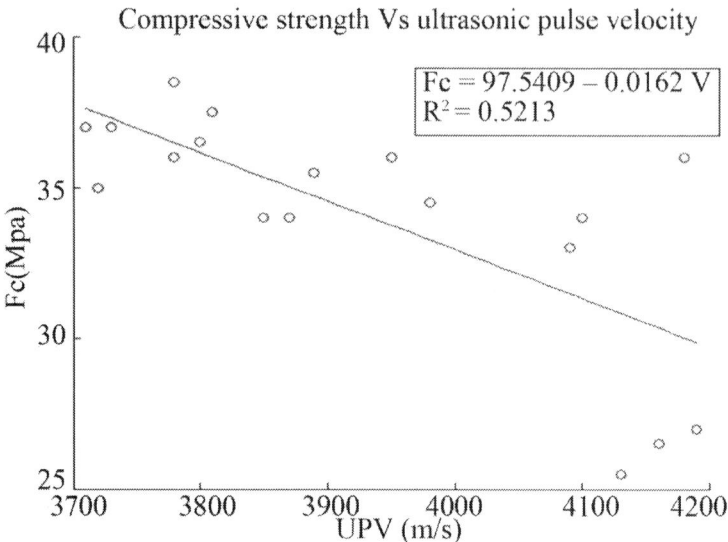

Figure 2. Correlation: compressive strength-ultrasonic pulse velocity (cylindrical specimens).

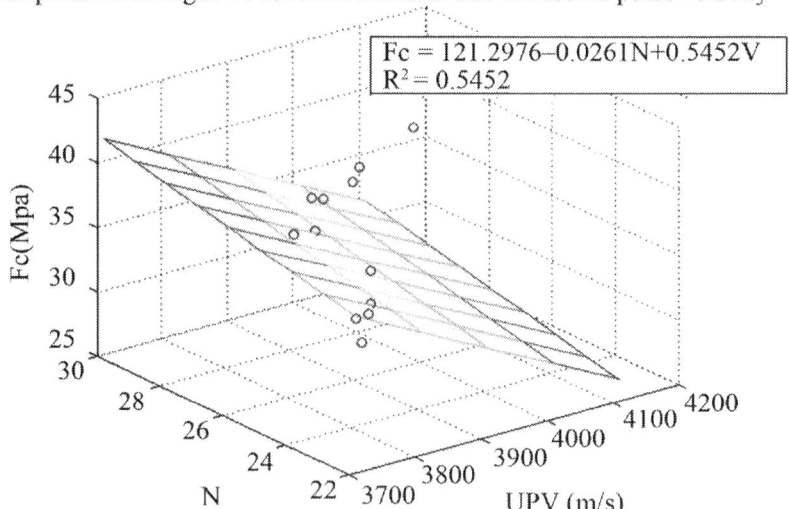

Figure 3. Correlation: compressive strength-rebound number and ultrasonic pulse velocity (cylindrical specimens).

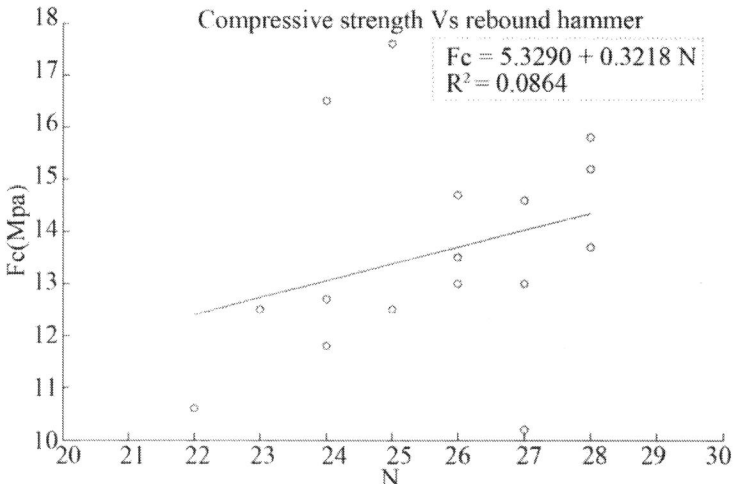

Figure 4. Correlation: compressive strength-rebound number (cores).

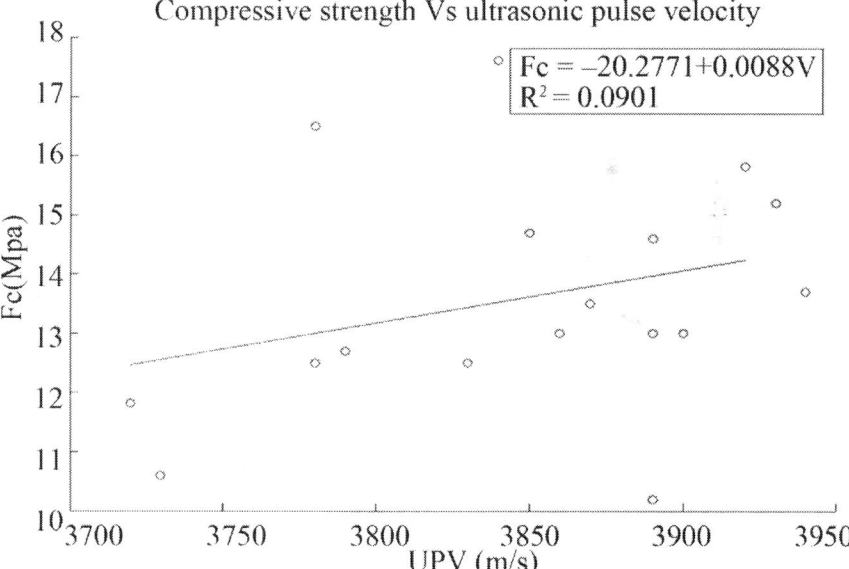

Figure 5. Correlation: compressive strength-ultrasonic pulse velocity (cores).

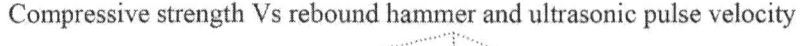

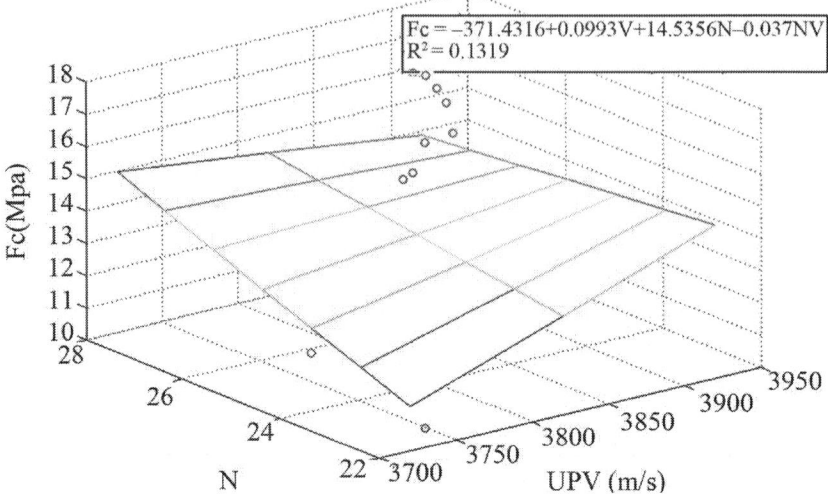

Figure 6. Correlation: compressive strength-rebound number and ultrasonic pulse velocity (cores).

Table 4. Regressions equations for cylindrical specimens.

Equations of régression and coefficient of détermination compressive strength (fc-MPa) cylindrical specimen	
Rebound hammer sclerométre N	$fc = -0.7708N + 54.6389$ $\qquad R^2 = 0.3983$
UPV	$fc = -0.0162V + 97.54095$ $\qquad R^2 = 0.5213$
Combined method	$fc = 0.5752V - 0.0261N + 121.2976$ $\quad R^2 = 0.5452$

Table 5. Regression equations for cores.

Equations of régression and coefficients of détermination for compressive strength (fc-MPa) of cores	
Rebound hammer	$fc = 0.3218N + 5.3290$ $\qquad R^2 = 0.0864$
UPV	$fc = 0.0088V - 20.2771$ $\qquad R^2 = 0.0901$
Combined method	$fc = 0.0993V + 14.5356N - 0.0037NV - 371.4$ $\quad R^2 = 0.1251$

The combined method seems more promising to evaluate the compressive strength of concrete in construction. It will also be noted that correlations between destructive testing and non-destructive techniques in our study provide more meaningful results for the specimens cast and stored under the same conditions as the concrete structure than taken cores.

The analysis for cores gives correlations that are not really satisfactory, this is explained by the fact that:

The quality and means of implementation of concrete which are often inadequate, in terms of social housing programs often attributed to small companies without major resources.

The sampling areas of taking cores are not really representative of concrete, since most often taken at random, because hardly feasible.

The core drilling way and conditions can affect the integrity of the cores.

Nevertheless these combined non-destructive methods can inform us about the quality of concrete and it will be better for a good quality-control monitoring of concrete; to establish correlations between mechanical tests on specimens cast and stored in same conditions as the concrete structure rather than using cores that are very difficult to achieve less representative and more expensive to obtain.

In general, the combined method appears more appropriate to conditions on-site measurements, very convenient, fast and with a reasonable cost. Once the correlations are established between the compressive strength values of samples derived from mechanical tests and measurements from non-destructive techniques (the rebound hammer and ultrasonic pulse velocity); the prediction of concrete strength value appears more reliable.

The practical use of this technique is gaining recognition on a large scale; it provides contracting authorities with accurate and objective information for monitoring quality-control of concrete construction.

REFERENCES

1. D. Breysse, "Quality of NDT Measurements and Accuracy of Physical Properties," Concrete NDTCE'09, Nantes, 30 June-3 July, 2009.

2. International Atomic Energy Agency, "Guidebook for the Fabrication of Nondestructive Testing Training Course," Series No. 13, 2001.

3. B. Hobbs and K. Tchoketch, "Nondestructive Testing Techniques for the Forensic Engineering Investigation of Reinforced Concrete Buildings," Forensic Science International, Vol. 167, No. 2-3, 2007, pp. 167-172. doi:10.1016/j.forsciint.2006.06.065

4. J. H. Bungey and S. G. Millard, "Testing of Concrete in Structures," 3rd Edition, Blackie Academic & Professional, London, 1996.

5. D. Breysse, "Combining Information, Reliability and Maintenance," Diagnosis of NEC Works, 2008

6. V. M. Malhotra and N. J. Carino, "CRC Handbook on Nondestructive Testing of Concrete," CRC Press, Boca Raton, 1991.

7. Rilem Report TC43-CND, "Draft Recommendation for in Situ Concrete Strength Determination by Combined NonDestructive Methods," 1983.

8. Y. Lin, C. Lai and T. Yen, "Prediction of Ultrasonic Pulse Velocity (UPV) in Concrete," ACI Materials Journal, Vol. 100, No. 1, 2003, pp. 21-28.

9. K. Komlos, S. Popovics, T. Nurnbergerova, B. Babal and J. S. Popovics, "Ultrasonic Pulse Velocity Test of Concrete Properties as Specified in Various Standard," Cement and Concrete Composites, Vol. 18, No. 5, 1996, pp. 357-364.

10. L. Divet, "Techniques for Diagnosing the Condition of Concrete," LCPC Presentation, Nantes, 2005.

11. I. Facaoaru, "Non-Destructive Testing of Concrete in Romania, Symposium on NDT of Concrete and Timber," Institute of Civil Engineers, London, 1970, pp. 39-49.

12. M. Erdal, "Prediction of the Compressive Strength of Vacuum Processed Concretes Using Artificial Neural Network and Regression Techniques," Scientific Research and Essay, Vol. 4, No. 10, 2009, pp. 1057-1065.

13. Hindo and Bergstrom, "Statistical Evaluation of the inPlace Strength of Concrete," Concrete International, Vol. 7, No. 2, 1985, pp. 44-48.

CITATION

S. Hannachi and M. Guetteche, "Application of the Combined Method for Evaluating the Compressive Strength of Concrete on Site," *Open Journal of Civil Engineering*, Vol. 2 No. 1, 2012, pp. 16-21. doi:10.4236/ojce.2012.21003.

CHAPTER 2

Analysis of the Optimum Usage of Slag for the Compressive Strength of Concrete

Han-Seung Lee [1], Xiao-Yong Wang [2,], Li-Na Zhang [2] and Kyung-Taek Koh [3]*

[1]Department of Architectural Engineering, Hanyang University, Ansan-Si 425-791, Korea;
[2]Department of Architectural Engineering, Kangwon National University, Chuncheon-Si 200-701, Korea;
[3]Structural Engineering Research Division, Korea Institute of Construction Technology, Goyang-Si 411-712, Korea;

ABSTRACT

Ground granulated blast furnace slag is widely used as a mineral admixture to replace partial Portland cement in the concrete industry. As the amount of slag increases, the late-age compressive strength of concrete mixtures increases. However, after an optimum point, any further increase in slag does not improve the late-age compressive strength. This optimum replacement ratio of slag is a crucial factor for its efficient use in the concrete industry. This paper proposes a numerical procedure to analyze the optimum usage of slag for the compressive strength of concrete. This numerical procedure starts with a blended hydration model that simulates cement hydration, slag reaction, and interactions between cement hydration and slag reaction. The amount of calcium silicate hydrate (CSH) is calculated considering the contributions from cement hydration and slag reaction. Then, by using the CSH contents, the compressive strength of the slag-blended concrete is evaluated. Finally, based on the parameter analysis of the compressive strength development of concrete with different slag inclusions, the optimum usage of slag in concrete mixtures is determined to be approximately 40% of the total binder content. The proposed model is verified through experimental results of the compressive strength of slag-blended concrete with different water-to-binder ratios and different slag inclusions.

INTRODUCTION

Slag is a by-product obtained during steel manufacturing and is commonly used in concrete because it improves durability and reduces porosity by improving the interface with the aggregate. Economic and ecologic benefits in the form of energy-savings and resource-conserving properties can also be achieved using slag-blended cement [1,2].

Compressive strength is the most important engineering property of concrete. To ensure progress in construction and safety in engineering practices, we aim to develop understanding on the strength development of concrete. Many experimental investigations have been conducted on the strength development of slag-blended concrete. The strength development of slag-blended concrete closely relates to the water-to-binder ratio, slag replacement ratio, and curing conditions. Beushausen *et al.* [3] found that, under moist curing conditions and when the slag replacement ratio is less than 50%, the 1-day early-age strength of concrete almost linearly decreases with the increase in the slag replacement ratio. At the ages of 28 and 56 days, due to the formation of calcium silicate hydrate (CSH) from the slag reaction, the compressive strength of slag-blended concrete can surpass that of control Portland cement concrete. Shariq *et al.* [4] found that, for concrete incorporating larger slag content (higher than 60% of the binder content), until the age of 180 days, the compressive strength of slag-blended concrete is still lower than that of Portland cement concrete. Oner and Akyuz [5] systematically investigated the effect of slag inclusions on the compressive strength development of concrete. They found that at a late age of 365 days, compressive strength of concrete mixtures containing slag increases as the amount of slag increases. After an optimum point of slag, a further increase in slag no longer improves the compressive strength.
To theoretically deduce the optimum usage of slag, models for compressive strength development of slag-blended concrete are necessary. Compared with abundant experimental studies [3,4,5], theoretical analysis of the compressive strength development of slag-blended concrete is limited. For comparing the relative performance of various supplementary cementing materials (SCMs: silica fume, fly ash, slag, natural pozzolans, *etc.*) as regards Portland cement, Papadakis [6,7] proposed an efficiency factor of SCMs that can be considered as equivalent to Portland cement. However, it should be noted that Papadakis' model [6,7] does not consider the effects of curing age and slag replacement ratios on the efficiency factor. Using a blended

hydration model considering both cement hydration and slag reaction, De Schutter [8,9] evaluated the early-age strength development of hardening slag-blended concrete. For late-age concrete, due to the significantly decreasing heat evolution rate, it is difficult to use De Schutter's model to evaluate the degree of hydration and strength development [2]. In addition, when the slag replacement ratio and water-to-binder ratio change, the coefficients of De Schutter's model will vary [2]. Using an artificial neural network, Bilim [10] evaluated the early-age strength and late-age strength of slag-blended concrete with different water-to-binder ratios and slag replacement ratios. However, we should note that artificial neural networks are a type of numerical regression method. Many parameters are necessary to build the input layer and hidden layers of artificial neural networks. The physical meaning of these parameters is not clear. Hence, it is difficult to adopt current models [5,6,7,8,9,10] to evaluate the strength development of slag-blended concrete with different mixing proportions. Moreover, current models [5,6,7,8,9,10] cannot be used to analyze the optimum usage of slag in concrete mixtures.

To overcome the weaknesses of the current research [5,6,7,8,9,10], this paper presents a numerical procedure to simulate the cement hydration, slag reaction, microstructure and strength development of hardening slag-blended concrete. The properties of concrete are determined considering contributions from cement hydration and slag reaction. Using parameter analysis of the compressive strength development of concrete with different slag inclusions, the optimum usage of slag in concrete mixtures is determined.

The innovations of this research are as follows. First, the proposed numerical procedure is valid for concrete with various mixing proportions, such as different water-to-binder ratios or different slag replacement ratios. The dependences of cement and slag reactivity on concrete mixing proportions and curing conditions are clarified; Second, the proposed numerical procedure is valid for both early-age concrete and late-age concrete. Evolutions of concrete properties are expressed as functions of reaction degrees of cement and slag; Third, the proposed numerical procedure evaluates the macro properties, such as the compressive strength of concrete, by using the microstructures of concrete such as the CSH content and phase volume fraction. The physical meanings of parameters in the proposed model are much clearer than those used in the artificial neural network model [10].

HYDRATION MODEL OF CEMENT–SLAG BLENDS

Hydration Model of Portland Cement

Tomosawa [11] proposed a shrinking-core model for the hydration of Portland cement. This model is expressed as a single equation consisting of three coefficients: kd, the reaction coefficient in the induction period; De, the effective diffusion coefficient of water through the C–S–H gel; and kri, a coefficient of the reaction rate of the mineral compound i of cement, as shown in Equations (1-1) and (1-2) below:

$$\frac{d\alpha_i}{dt} = \frac{3\left(S_w/S_0\right)\rho_w C_{w-free}}{(v+w_g)r_0\rho_c} \cdot \frac{1}{\left(\frac{1}{k_d} - \frac{r_0}{D_e}\right) + \frac{r_0}{D_e}(1-\alpha_i)^{\frac{-1}{3}} + \frac{1}{k_{ri}}(1-\alpha_i)^{\frac{-2}{3}}}$$

(1-1)

$$\alpha = \frac{\sum_{i=1}^{4}\alpha_i g_i}{\sum_{i=1}^{4} g_i}$$

(1-2)

where αi (i = 1, 2, 3, and 4) represents the reaction degree of the cement mineral compounds C_3S, C_2S, C_3A, and C_4AF, respectively; α is the degree of cement hydration and can be calculated from the weight fraction of mineral compound gi and reaction degree of mineral compound αi; v is the stoichiometric ratio of the mass of water to cement (=0.25); wg is the physically bound water in C–S–H gel (=0.15); ρw is the density of water; ρc is the density of the cement; Cw−free is the amount of water at the exterior of the C–S–H gel; r0 is the radius of unhydrated cement particles; Sw is the effective surface area of the cement particles in contact with water; and S0 is the total surface area if the surface area develops unconstrained.

The reaction coefficient kd is assumed to be a function of the degree of hydration, as shown in Equation (2), where B and C are the coefficients determining this factor; Bcontrols the rate of the initial shell formation, and C controls the rate of the initial shell decay.

$$k_d = \frac{B}{\alpha^{1.5}} + C\alpha^3$$

(2)

The effective diffusion coefficient of water is affected by the tortuosity of the gel pores as well as the radii of the gel pores in the hydrate. This phenomenon can be described as a function of the degree of hydration and is expressed as follows:

$$D_e = D_{e0} \ln(\frac{1}{\alpha}) \qquad (3)$$

In addition, free water in the capillary pores is depleted as hydration of cement minerals progresses. Some water is bound in the gel pores, and this water is not available for further hydration, an effect that must be taken into consideration in every step of the progress of the hydration. Therefore, the amount of water in the capillary pores, Cw–free, is expressed as a function of the degree of hydration in the previous step, as shown in Equation (4).

$$C_{w-free} = \frac{W_0 - 0.4 * \alpha * C_0}{W_0} \qquad (4)$$

where C_0 and W_0 are the mass fractions of cement and water in the mix proportion.

The effect of temperature on these reaction coefficients is assumed to follow Arrhenius's law as shown in Equations (5)–(8):

$$B = B_{20} \exp(-\beta_1(\frac{1}{T} - \frac{1}{293})) \qquad (5)$$

$$C = C_{20} \exp(-\beta_2(\frac{1}{T} - \frac{1}{293})) \qquad (6)$$

$$k_{ri} = k_{ri20} \exp(-\frac{E}{R}(\frac{1}{T} - \frac{1}{293})) \qquad (7)$$

$$D_e = D_{e20} \exp(-\beta_3(\frac{1}{T} - \frac{1}{293})) \tag{8}$$

where β_1, β_2, E/R, and β_3 are temperature sensitivity coefficients and B_{20}, C_{20}, kri_{20}, and De_{20} are the values of B, C, kri, and De at 20 °C.

On the basis of the degree of reaction of the mineral compounds of cement [12], the parameters of the hydration model are calibrated and shown in Table 1. Using this Portland cement hydration model, Tomosawa [11] evaluated the heat evolution rate, adiabatic temperature rise, compressive strength development, and thermal stress development in both ordinary strength concrete and high strength concrete. However, it should be noted that Tomosawa's model is valid only for Portland cement. For slag-blended cement, due to the coexistence of Portland cement hydration and the chemical reaction of slag, Tomosawa's model is not valid. To model the hydration of slag-blended concrete, the reaction model of slag should be built and the mutual interactions between cement hydration and slag reaction should be clarified.

Table 1. Coefficients of the cement hydration model.

B_{20} (cm/h)	C_{20} (cm/h)	k_{rC_3S20} (cm/h)	k_{rC_2S20} (cm/h)	k_{rC_3A20} (cm/h)	k_{rC_4AF20} (cm/h)	D_{e20} (cm²/h)	β_1 (K)	β_2 (K)	β_3 (K)	$\frac{E}{R}$ (K)
8.09×10^{-9}	0.02	9.03×10^{-6}	2.71×10^{-7}	1.35×10^{-6}	6.77×10^{-8}	8.62×10^{-10}	1000	1000	7500	5400

Slag Reaction Model

The hydration rate of slag depends on the amount of calcium hydroxide in the hydrating cement-slag blends and the reaction degree of the mineral admixtures [8,11,12,13]. Compared with silica fume, the hydration rate of slag is much lower due to the larger particle size. Similar to the hydration process of cement, the reaction of slag can be divided into three processes: an initial dormant period, the phase-boundary reaction and diffusion processes [8,13]. By considering these points, we originally proposed that the reaction equation of slag can be written as follows:

$$\frac{d\alpha_{SG}}{dt}$$
$$= \frac{CH(t)}{P} \frac{W_{cap}}{W_0} \frac{3\rho_w}{\nu_{SG} r_{SG0} \rho_{SG}} \frac{1}{(\frac{1}{k_{dSG}} - \frac{r_{SG0}}{D_{eSG}}) + \frac{r_{SG0}}{D_{eSG}}(1 - \alpha_{SG})^{-\frac{1}{3}} + \frac{1}{k_{rSG}}(1 - \alpha_{SG})^{-\frac{2}{3}}}$$

$$\tag{9-1}$$

$$k_{dSG} = \frac{B_{SG}}{(\alpha_{SG})^{1.5}} + C_{SG} * (\alpha_{SG})^3$$

$$(9\text{-}2)$$

$$D_{eSG} = D_{eSG0} * \ln\left(\frac{1}{\alpha_{SG}}\right) \qquad (9\text{-}3)$$

where α_{SG} is the degree of the reaction of slag; CH(t) is the calcium hydroxide mass in a unit volume of hydrating cement-slag blends; P is the mass of slag in the mixture proportion; Wcap is the mass of capillary water; vSG is the stoichiometric ratio of the mass of CH to slag; rSG0 is the radius of a slag particle; ρSG is the density of the slag; kdSG is the reaction rate coefficient in the dormant period (BSG and CSG are coefficients); DeSG0 is the initial diffusion coefficient; and krSG is the reaction rate coefficient. Slag shows both cementitious behavior (latent hydraulic activity) and pozzolanic characteristics (reaction with lime). In Equation (9-1), the term CH(t)P considers the pozzolanic characteristics of slag, and the term WcapW0 considers the latent hydraulic activity of slag.

The influence of temperature on the slag reaction is originally considered by the Arrhenius law as follows:

$$B_{SG} = B_{SG20} \exp\left(-\beta_{1SG}\left(\frac{1}{T} - \frac{1}{293}\right)\right) \qquad (9\text{-}4)$$

$$C_{SG} = C_{SG20} \exp\left(-\beta_{2SG}\left(\frac{1}{T} - \frac{1}{293}\right)\right) \qquad (9\text{-}5)$$

$$D_{eSG0} = D_{eSG20} \exp\left(-\beta_{3SG}\left(\frac{1}{T} - \frac{1}{293}\right)\right)$$

$$(9\text{-}6)$$

$$k_{rSG} = k_{rSG20} \exp(-\frac{E_{SG}}{R}(\frac{1}{T} - \frac{1}{293}))$$

(9-7)

where B_{SG20}, C_{SG20}, D_{eSG20}, and k_{rSG20} are the values of B_{SG}, C_{SG}, D_{eSG0}, and k_{rSG} at 293 K, respectively, and β_{1SG}, β_{2SG}, β_{3SG}, and E_{SG}/R are the temperature sensitivity coefficients of B_{SG}, C_{SG}, D_{eSG0}, and k_{rSG}, respectively. The temperature sensitivity coefficients of slag can be determined from the reaction degree of slag at different curing temperatures [11,14].

Interactions between Cement Hydration and Slag Reaction

Based on analysis of the experimental results of the amount of chemically bound water, adiabatic temperature rise, and temperature measurement of small quasi-adiabatic blocks, Maekawa *et al.* [14] stated that the reaction of slag can be roughly described by the following approximate key figures:

Calcium hydroxide	0.22 g/g slag
Chemically bound water	0.30 g/g slag
Gel water	0.15 g/g slag

Using the hydration degree of cement, reaction degree of slag, and stoichiometry of the reaction of slag [14], the amounts of calcium hydroxide, chemically bound water, and capillary water in cement-slag blends during hydration can be originally determined with the following equations:

$$CH(t) = RCH_{CE} * C_0 * \alpha - 0.22 * \alpha_{SG} * P$$

(10)

$$W_{cap} = W_0 - 0.4 * C_0 * \alpha - 0.30 * \alpha_{SG} * P - 0.15 * \alpha_{SG} * P$$

(11)

$$W_{cbm} = v * C_0 * \alpha + 0.3 * \alpha_{SG} * P$$

(12)

In Equation (10), RCHCE is the mass of produced calcium hydroxide from the hydration of cement. In Equation (12), Wcbm is the mass of chemically bound water. As shown in Equations (10)–(12), the evolution of calcium hydroxide, chemically bound water, and capillary water in cement-slag blends depends on both cement hydration and slag reaction.

As proposed by Papadakis [6,7], for slag-blended concrete, the calcium silicate hydrate (CSH) content, which is the most critical parameter in strength development, can be calculated as a function of the cement content, the slag content, the weight fraction of silica in cement fS,C and slag fS,P, and the weight fraction of the reactive oxide SiO_2 in the slag γS. Combining Papadakis' chemical reaction equation [6,7] and the hydration reaction Equations (1) and (9), the amount of CSH in hardening slag-blended concrete can be initially calculated as follows:

$$CSH(t) = 2.85(f_{S,C} * C_0 * \alpha + \gamma_s * f_{S,P} * P * \alpha_{SG}) \quad (13)$$

When slag is incorporated into concrete, two possible reasons may be adopted to explain the change in the hydration process. One is the chemical reaction of amorphous phases in slag, and the other is the influence of slag on the hydration of cement. In the current paper, the new model that is originally proposed can describe the reaction of slag. The influence of slag on the hydration of cement is originally considered through the amount of capillary water (Equation (11)) and the dilution effect (Equation (4)) [14]. Hence, the proposed model shows a strong ability to simulate the hydration of concrete containing slag. Furthermore, the development of properties of hardening slag-blended concrete can be evaluated based on the degree of the reactions of cement and slag.

Calibration of the Reaction Coefficients of the Slag Reaction Model

Iyoda *et al.* [15] investigated the reaction degrees of slag in cement-slag paste considering the effects of curing temperatures (5, 20 and 40 °C) and slag replacement ratios (42% and 67% mass fractions). The water-to-binder ratio of cement-slag paste is 0.5. At the ages of 1, 3, 7, 28, 56, and 91 days, the reaction degree of slag was measured using a selective dissolution method. The chemically bound water was measured using an ignition loss method.

For each curing temperature, 5, 20 and 40 °C, the reaction coefficients of slag, BSG, CSG,DeSG, and krSG, can be calibrated through experimental results of the reaction degrees of slag. Furthermore, the temperature sensitivity coefficients of BSG, CSG, DeSG, and krSGcan be determined using the reaction coefficients at different curing temperatures. The values of the slag reaction coefficients and the temperature sensitivity coefficients are shown in Table 2. These fitted parameters for slag are not changed from one mix to another. When the water-to-binder ratio or slag replacement is changed, these parameters of slag do not vary.

Table 2. Coefficients of the slag reaction model.

B_{20SG} (cm/h)	$CSG20$ (cm/h)	k_{rSG20} (cm/h)	D_{eSG20} (cm²/h)	β_{1SG} (K)	β_{2SG} (K)	β_{3SG} (K)	$\dfrac{E_{SG}}{R}$ (K)
8.93×10^{-9}	0.1	1.0×10^{-5}	1.86×10^{-9}	1000	1000	5000	7000

As shown in Figure 1, with the increase in curing temperature, the reaction degree of slag increases correspondingly. In addition, when the replacement ratio of slag increases from 42% (Figure 1a) to 67% (Figure 1b), due to the shortage of calcium hydroxide, the reaction degrees of slag will decrease significantly. As shown in Figure 2, due to the increasing curing temperatures, the cement hydration and slag reaction will accelerate. Given a certain age, more chemically bound water will be produced for paste with a higher curing temperature. For cement-slag paste cured at 40 °C, the increment of chemically bound water content is marginal after the age of 28 days because of the decreasing reaction rate of cement and slag.

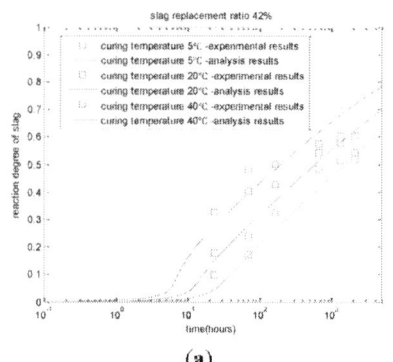

(a)

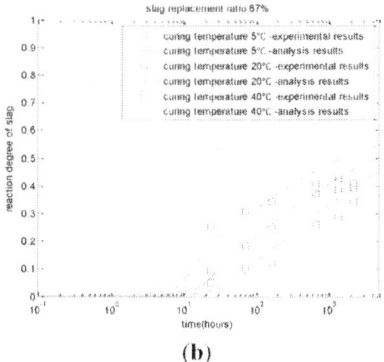

(b)

Figure 1. Reaction degree of slag with different slag replacement ratios and curing temperatures (experimental results are taken from reference [15]). (**a**) cement-slag paste with 42% slag; (**b**) cement-slag paste with 67% slag.

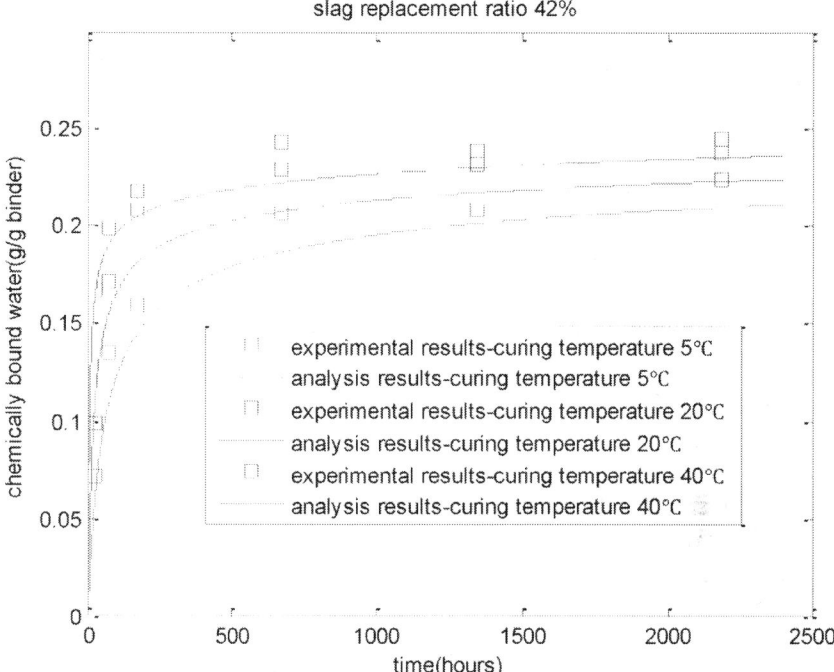

Figure 2. Evaluation of chemically bound water of cement-slag blends (water-to-binder ratio of 0.42) (experimental results are taken from reference [15]).

The flow chart of modeling is summarized in Figure 3. At each time step, the cement hydration degree and slag reaction degree are calculated. The calcium hydroxide contents, capillary water contents, and chemically bound water contents are determined considering contributions from cement hydration and slag reaction. Furthermore, by using the reaction degrees of cement and slag, the phase volume fractions, calcium silicate hydrate content [16], and strength development of hardening concrete can be determined.

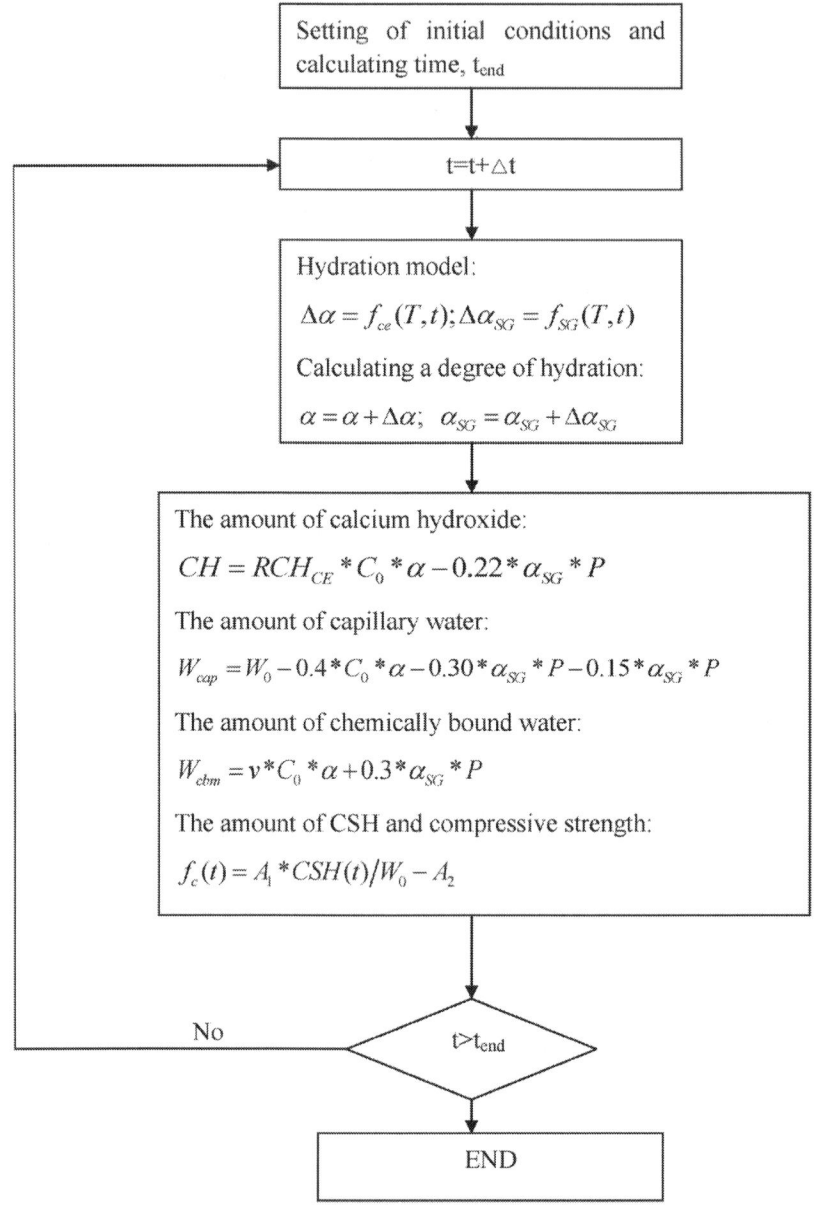

Figure 3. Flowchart of modeling.

ANALYSIS OF OPTIMUM USAGE OF SLAG FOR THE COMPRESSIVE STRENGTH OF CONCRETE

As reported by Papadakis [6,7], the compressive strength of concrete is mainly dependent on the amount of calcium silicate hydrate (CSH). For hardening slag-blended concrete, as shown in Equation (13), the amount of CSH relates to both the cement hydration and slag reaction. As proposed by Neville [17], the relation between the compressive strength of concrete and CSH contents can be described by a linear equation as follows:

$$f_c(t) = A_1 * \frac{CSH(t)}{W_0} - A_2 \tag{14}$$

where $f_c(t)$ is the compressive strength of concrete, and A1 and A2 are coefficients related to compressive strength. As shown in this equation, for hardening concrete, the compressive strength development starts after a threshold degree of hydration. When the degree of hydration is less than this threshold degree of hydration, the compressive strength of concrete is zero. The concept of threshold degree of hydration is similar to that of the final setting time of concrete (final set means complete solidification and the beginning of hardening, and in concrete technology, hardening is the phenomenon of strength gain with time [1]).

The experimental results of Cheng *et al.* [18] are used to verify the proposed model. In the study, three water/binder ratios (0.35, 0.50, and 0.70) and three substitution ratios of cement with slag (10%, 20%, and 40%) were selected for the preparation of concrete specimens. The cement used is ASTM type I cement, the fineness of the granulated blast furnace slag is 4000 cm^2/g, the maximum size of the coarse aggregate is 20 mm, and the fineness modulus of the fine aggregate is 2.96. The chemical compositions of cement and slag are shown in Table 3 and Table 4, respectively. The mixture proportions of concrete are shown in Table 5. The size of the cylinder specimen for the compression test is 100 × 200 mm. The specimens were tested at five ages (1, 3, 7, 28, and 56 days) for the compressive strength measurement.

Using the blended cement hydration model, the amount of CSH can be calculated and is shown in Figure 4. As shown in this figure, for early-age slag-blended concrete, the produced CSH contents are less than that of control concrete, while for late-age slag-blended concrete, the CSH contents can surpass that of the control Portland cement concrete. With the increase in the slag replacement ratio from 0.2 to 0.6, the age corresponding to the surpassing of CSH will be postponed. Alternately, based on the calculated CSH contents and measured compressive strength of concrete, the strength coefficients of Equation (14) can be calibrated. The values of A_1 for a water-to-binder ratio of 0.7, 0.5, and 0.35 are 52.39, 61.54, and 55.37, respectively. The values of A_2 for a water-to-binder ratio of 0.7, 0.5, and 0.35 are 9.81, 12.04, and 4.47, respectively.

Table 3. Chemical composition of Portland cement [18].

Chemical Composition (mass %)							Blaine (cm^2/g)
SiO_2	Al_2O_3	Fe_2O_3	CaO	MgO	SO_3	L.O.I	
20.6	4.0	6.1	62.8	2.6	3.1	0.8	3090

Table 4. Chemical composition of slag [18].

Chemical Composition (mass %)									Blaine (cm^2/g)
SiO_2	Al_2O_3	Fe_2O_3	CaO	MgO	SO_3	Na_2O	K_2O	L.O.I	
34.4	9.0	2.58	44.8	4.43	2.26	0.62	0.5	1.32	4000

Table 5. Mixing proportions of concrete containing slag [18].

	Water-to-Binder Ratio	Slag Replacement Ratio	Water (kg/m³)	Cement (kg/m³)	Slag (kg/m³)	Sand (kg/m³)	Aggregate (kg/m³)	Water Reducing Agent (Binder ×%)
WB35	0.35	-	202.8	591	0	570	973	4.1
WB35-10	0.35	10%	202.8	532	59	565	973	4.1
WB35-20	0.35	20%	202.8	473	118	560	973	4.1
WB35-40	0.35	40%	202.8	355	236	552	973	4.1
WB50	0.5	-	206.5	414	0	718	973	0.4
WB50-10	0.5	10%	206.5	372	41	715	973	0.4
WB50-20	0.5	20%	206.5	331	83	712	973	0.4
WB50-40	0.5	40%	206.5	248	165	706	973	0.4
WB70	0.7	-	206.9	296	0	815	973	0
WB70-10	0.7	10%	206.9	266	29	815	973	0
WB70-20	0.7	20%	206.9	237	59	812	973	0
WB70-40	0.7	40%	206.9	177	118	807	973	0

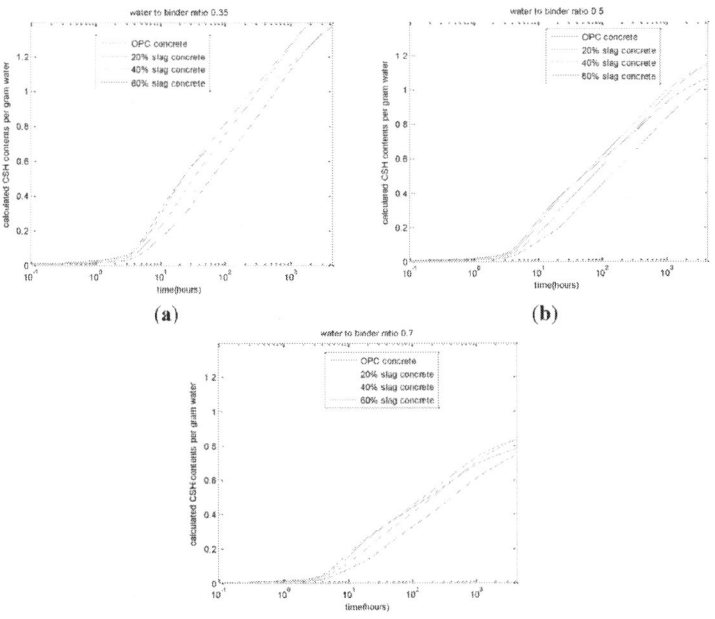

(a)

(b)

(c)

Figure 4. Calculated calcium silicate hydrate (CSH) content. (**a**) water-to-binder ratio of 0.35; (**b**) water-to-binder ratio of 0.5; (**c**) water-to-binder ratio of 0.7.

At the macroscopic level, concrete is a composite material consisting of discrete aggregates dispersed in a continuous cement paste matrix. The bonding region or interfacial transition zone (ITZ) in concrete between the matrix and aggregate is a critical component of mechanical performance [19]. For ordinary strength concrete (water-to-binder ratio of 0.5) and low strength concrete (water-to-binder ratio of 0.7), the ITZ is the weak link of concrete, and the compressive strength of concrete is mainly dependent on the strength of the ITZ. When the water-to-cement ratio decreases from 0.7 (low strength concrete) to 0.5 (ordinary strength concrete), the distribution of reaction products in the ITZ becomes more homogeneous [16], and the contribution of 1 gram of CSH to the compressive strength will increase. Hence, when the water-to-cement ratio decreases from 0.7 to 0.5, the strength coefficient A_1 increases from 52.39 to 61.54. Alternately, for high strength concrete (water-to-binder ratio of 0.35), the strength of concrete relates to three phases of concrete, *i.e.*, the ITZ phase, bulk paste matrix phase, and aggregate phase. Due to the contribution of the aggregate to the compressive strength, the ratio of the strength of the ITZ to the sum of the other two phases (bulk paste matrix paste plus aggregate phase) will decrease. Thus, when the water-to-cement ratio decreases from 0.5 to 0.3, the strength coefficient A_1 also decreases from 61.54 to 55.37.

Figure 5 shows the analysis results for the compressive strength development of slag-blended concrete. First, the proposed model can reflect the effect of the water-to-cement ratio on the compressive strength development of concrete. With the increase in the water-to-cement ratio, given a certain age, the produced CSH content for 1 gram of mixing water will decrease. Hence, the compressive strength will decrease correspondingly; Second, as shown in Figure 5a,c,e, the proposed model can reproduce the strength crossover phenomenon between the control Portland cement concrete and the slag-blended concrete. Because the reactivity of slag is lower than that of Portland cement, the early-age strength of slag-blended concrete is lower than that of the control concrete. However, because the produced CSH content from 1 gram of reacted slag is higher than that from 1 gram of hydrated cement, at late age, for concrete containing 10%, 20%, and 40% slag, the compressive strength of slag-blended concrete can surpass that of the control concrete. In addition, with the increase in the slag replacement ratio, the reactivity of slag will decrease and the age corresponding to the crossover of the compressive strength will be postponed; Third, as shown in Figure 5a,c,e, given certain water-to-binder ratios, with the increase in slag content, the X-axis intercept of the strength development function increases correspondingly. This agrees with Brooks *et al.*'s study [20] on the setting time of slag-blended concrete. They reported that the inclusion of slag at replacement levels of 40% and greater resulted in significant retardation in setting times. As the replacement levels of slag were increased, there was greater retardation in setting times. On the other hand, as shown in Figure 5b,d,f, with the increase in the water-to-binder ratio, the X-axis intercept of the strength development function will increase. Neville [18] also reported that the setting of concrete increases with an increase in the water-to-binder ratio.

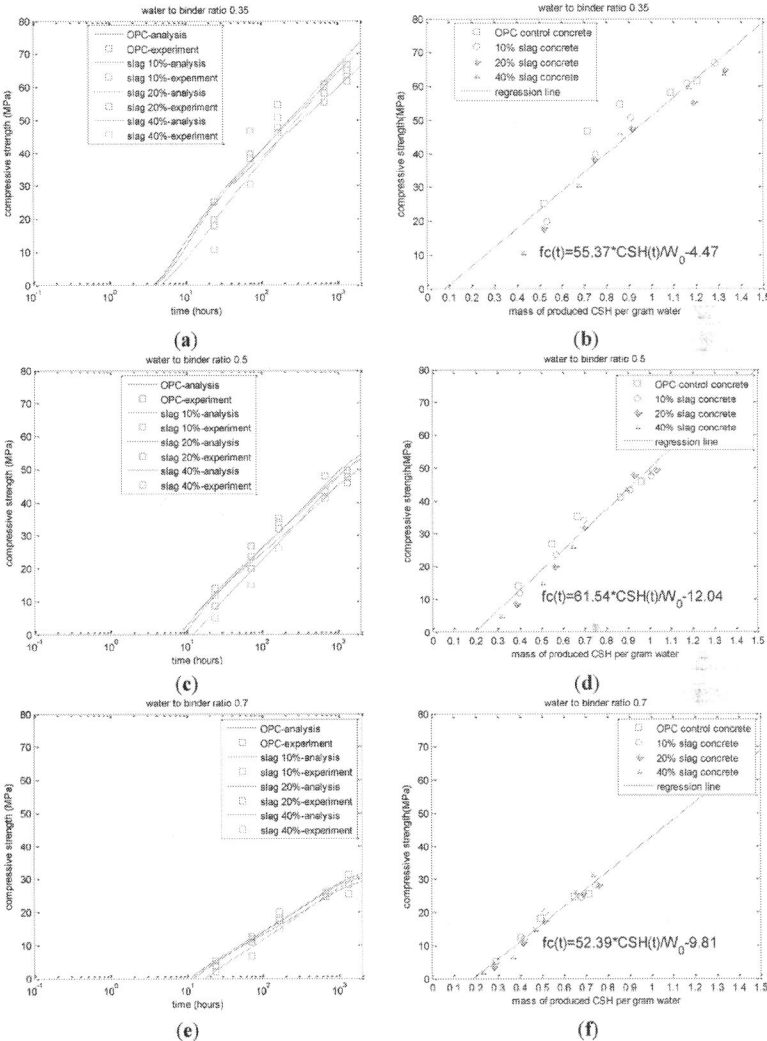

Figure 5. Analysis of the compressive strength development (experimental results are taken from reference [18]). (**a**) compressive strength *versus* age: water-to-binder ratio of 0.35; (**b**) compressive strength*versus* CSH: water-to-binder ratio of 0.35; (**c**) compressive strength *versus* age: water-to-binder ratio of 0.5; (**d**) compressive strength *versus* CSH: water-to-binder ratio of 0.5; (**e**) compressive strength *versus* age: water-to-binder ratio of 0.7; (**f**) compressive strength*versus* CSH: water-to-binder ratio of 0.5.

Figure 6 presents a holistic comparison between the experimental results and the analysis results. As shown in this figure, the analysis results generally agree with the experimental results. For high strength concrete with a water-to-binder ratio of 0.35, due to the ignorance of the aggregate contribution to the compressive strength, the analysis results slightly deviate from the experimental results.

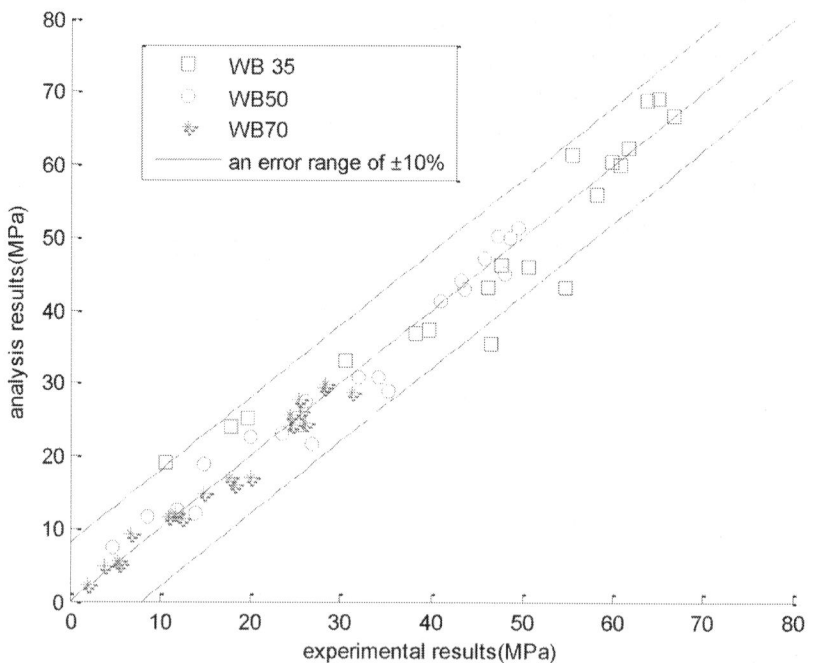

Figure 6. Comparison between the analysis results and experimental results (experimental results are taken from reference [18]).

Figure 7 presents the evolution of the phase volume fractions of hardening cement-slag blends (water-to-binder ratio of 0.5 with 50% slag, curing temperature of 20 °C). As shown in this figure, with the progression of cement hydration and slag reaction, the volumes of un-reacted cement and slag decrease, the volumes of CSH and other reaction products increase, and due to the filling effects of the reaction products, the volume of the capillary pore decreases. At an early age, cement hydration and slag reaction proceed quickly, and at a late age, the reaction rates become slower. Because the reactivity of cement is much higher than that of slag, at the age of 180 days, the volume of anhydrous cement is much less than that of unreacted slag.

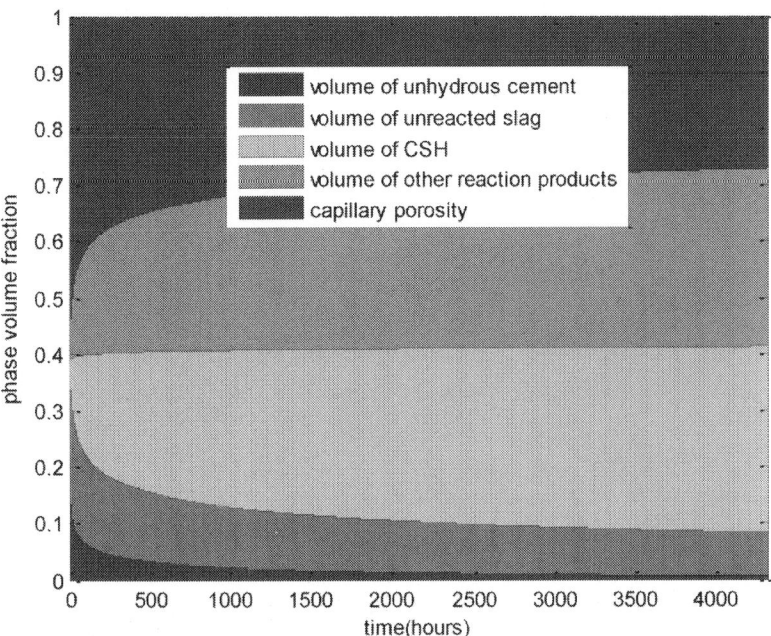

Figure 7. Phase volume fraction evolutions of cement-slag paste: water-to-binder ratio of 0.5, slag replacement ratio of 0.5.

Figure 8 shows the parameter analysis of the effect of slag inclusion on the compressive strength development of concrete. The water-to-binder ratios shown in Figure 8a–c are 0.35, 0.5, and 0.7, respectively. The vertical axis of these figures represents the ratio of the compressive strength between slag-blended concrete and the control Portland cement concrete. As shown in the figures, at the early age of 1 day, with the increase in the slag replacement ratio, the compressive strength of slag-blended concrete almost linearly decreases. As the curing age increases, obviously, the strength of slag-blended concrete with higher slag ratios increases faster, and at a late age, such as 360 days, the maximum value of the strength lies roughly at the slag replacement ratio of 40%. With regard to a slag replacement ratio higher than 40%, due to the lower reactivity of slag (shown in Figure 1), the ultimate strength ratio is less. At the age of 360 days, because of losses in the capillary water, the decrease in available deposition spaces for hydration products, and a change in the hydration rate-determining process to a diffusion-controlled stage, the rate of the reaction becomes much slower, and the change in the compressive strength ratio between slag-blended concrete and OPC concrete is very marginal. Based on the evolution of the compressive strength ratio of concrete with different water-to-binder ratios and different slag inclusions, a slag replacement ratio of 40% can be regarded as the optimum slag content of concrete.

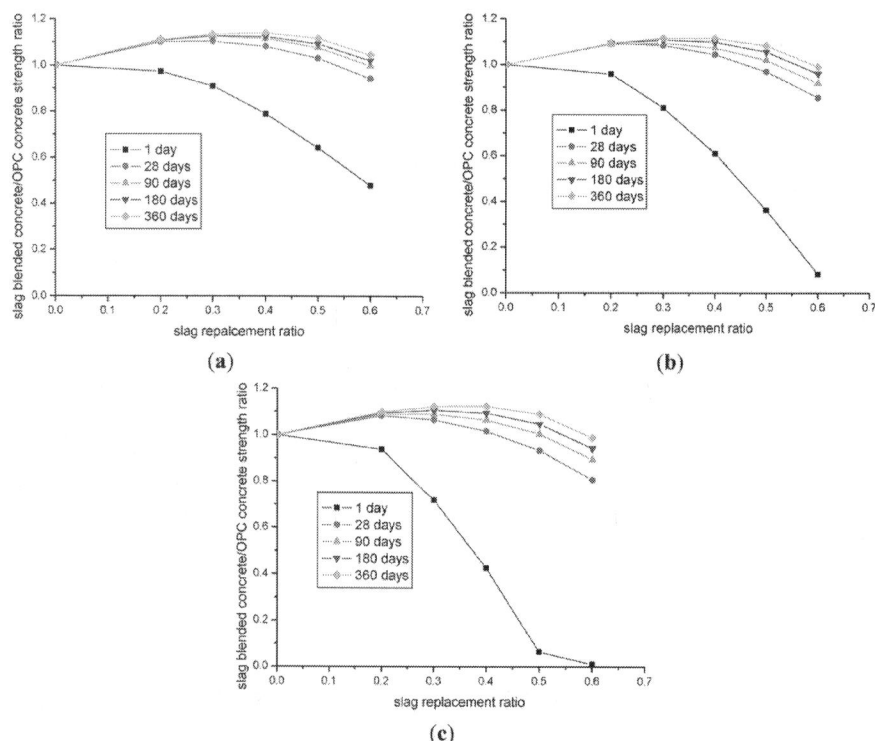

Figure 8. Effects of slag inclusions on the compressive strength development of concrete. (**a**) water-to-binder ratio of 0.35; (**b**) water-to-binder ratio of 0.5; (**c**) water-to-binder ratio of 0.7.

A summary regarding the determination of optimum slag content is as follows: First, using experimental results on the reaction degree of slag, the reaction coefficients of the slag reaction model are calibrated; Second, using experimental results on the compressive strength, the strength development coefficients of slag-blended concrete are calibrated; Third, parameter analysis of the compressive strength development of Portland cement concrete and slag-blended concrete is performed. By comparing the strength development of slag-blended concrete with that of Portland cement concrete, the optimum slag content can be analyzed.

CONCLUSIONS

This paper has proposed a numerical procedure to analyze the optimum usage of slag for the compressive strength of concrete. The conclusions arc summarized as follows.

First, we proposed a blended hydration model that simulates cement hydration, slag reaction, interactions between cement hydration and slag reaction, and phase volume fraction evolution of cement–slag blends. The amount of calcium silicate hydrate (CSH), which is closely related to the compressive strength of concrete, is calculated considering the contributions from both cement hydration and slag reaction.

Second, by using a linear equation for the compressive strength and CSH content, the development of the strength of slag-blended concrete is evaluated. Given a certain age, with an increase in the water-to-binder ratio, the compressive strength of concrete will decrease. The proposed model can reproduce the strength crossover phenomenon between the control Portland cement concrete and slag-blended concrete. The early-age strength of slag-blended concrete is lower than that of the control concrete. However, at a late age, for concrete containing 10%, 20%, and 40% slag, the compressive strength of slag-blended concrete can surpass that of the control concrete. With an increase in the slag content and the water-to-binder ratio, the starting age of the compressive strength development will be retarded.

Third, based on parameter analysis, the ratio of the compressive strength between slag-blended concrete and Portland cement concrete is calculated. For concrete with different water-to-binder ratios, at the late age of 360 days, the optimum slag content is approximately 40% of the total binder content. Before this optimum point, the compressive strength of concrete mixtures containing slag increases as the amount of slag increases. After an optimum point, any further increase in slag inclusion does not improve the compressive strength.

ACKNOWLEDGMENTS

This research was supported by a grant (13SCIPA02) from the Smart Civil Infrastructure Research Program funded by the Ministry of Land,

Infrastructure and Transport (MOLIT) of the Korean government and the Korea Agency for Infrastructure Technology Advancement (KAIA).

AUTHOR CONTRIBUTIONS

Han-Seung Lee conducted the programming and wrote the initial draft of the manuscript. Xiao-Yong Wang designed the project and analyzed the data. Li-Na Zhang and Kyung-Taek Koh wrote the final manuscript. All authors contributed to the analysis of the data and read the final paper.

REFERENCES

1. Metha, P.K.; Paulo, J.M. *Concrete, Microstructure, Properties and Materials*; McGraw-Hill: New York, NY, USA, 2006; pp. 281–317.
2. Wang, X.Y.; Lee, H.S. Modeling the hydration of concrete incorporating fly ash or slag. *Cem. Concr. Res.* 2010, *40*, 984–996.
3. Beushausen, H.; Alexander, M.; Ballim, Y. Early-age properties, strength development and heat of hydration of concrete containing various South African slags at different replacement ratios. *Constr. Build. Mater.* 2012, *29*, 533–540.
4. Shariq, M.; Prasad, J.; Masood, A. Effect of GGBFS on time dependent compressive strength of concrete. *Constr. Build. Mater.* 2010, *24*, 1469–1478.
5. Oner, A.; Akyuz, S. An experimental study on optimum usage of GGBS for the compressive strength of concrete. *Cem. Concr. Comp.* 2007, *29*, 505–514.
6. Papadakis, V.G.; Antiohos, S.; Tsimas, S. Supplementary cementing materials in concrete Part II: A fundamental estimation of the efficiency factor. *Cem. Concr. Res.* 2002, *32*, 1533–1538.
7. Sotiris, D.; Maria, P.E.; Papadakis, V.G. Computer-aided modeling of concrete service life. *Cem. Concr. Comp.* 2014, *47*, 9–18.
8. De Schutter, G. Finite element simulation of thermal cracking in massive hardening concrete elements using degree of hydration based material laws. *Comput. Struct.* 2002, *80*, 2035–2042.

9. Jiang, W.; de Schutter, G.; Yuan, Y. Degree of hydration based prediction of early age basic creep and creep recovery of blended concrete. *Cem. Concr. Comp.* 2014,*48*, 83–90.

10. Bilim, C.; Atis, C.D.; Tanyildizi, H.; Karahan, O. Predicting the compressive strength of ground granulated blast furnace slag concrete using artificial neural network. *Adv. Eng. Softw.* 2009, *40*, 334–340.

11. Tomosawa, F. Development of a kinetic model for hydration of cement. In Proceedings of the Tenth International Congress Chemistry of Cement, Gothenburg, Sweden, 2–6 June 1997; Chandra, S., Ed.; pp. 51–58.

12. Matsushita, T.; Hoshino, S.; Maruyama, I.; Noguchi, T.; Yamada, K. Effect of curing temperature and water to cement ratio on hydration of cement compounds. In Proceedings of the 12th International Congress on the Chemistry of Cement, Montreal, QC, Canada, 10–13 April 2007; Beaudoin, J., Ed.; .

13. Saeki, T.; Monteiro, P.J.M. A model to predict the amount of calcium hydroxide in concrete containing mineral admixture. *Cem. Concr. Res.* 2005, *35*, 1914–1921.

14. Maekawa, K.; Ishida, T.; Kishi, T. *Multi-Scale Modeling of Structural Concrete*; Taylor & Francis: London, UK; New York, NY, USA, 2009; pp. 46–106.

15. Iyoda, T.; Inokuchi, K.; Uomoto, T. Effect of slag hydration of blast furnace slag cement in different curing conditions. In Proceedings of the 13th International Congress on the Chemistry of Cement, Madrid, Spain, 3–8 July 2011.

16. Wang, X.Y.; Lee, H.S.; Park, K.B. Simulation of low-calcium fly ash blended cement hydration. *ACI Mater. J.* 2009, *106*, 167–175.

17. Neville, A.M. *Properties of Concrete*, 4th and final ed.; John Wiley & Sons. Inc.: Oxford, UK, 1996; pp. 269–318.

18. Cheng, A.S.; Yen, T.; Liu, Y.W.; Sheen, Y.N. Relation between porosity and compressive strength of slag concrete. In *Structures Congress*; Ventura, C., Hoit, M., Anderson, D., Harvey, D., Eds.; ASCE: Vancouver, BC, Canada, 2008.

19. Wang, X.Y.; Lee, H.S.; Park, K.B.; Kim, J.J.; Golden, J.S. A multi-phase kinetic model to simulate hydration of slag–cement blends. *Cem. Concr. Comp.* 2010, *32*, 468–477.

20. Brooks, J.J.; Johari, M.A.M.; Mazloom, M. Effect of admixtures on the setting times of high-strength concrete. *Cem. Concr. Comp.* 2000, *22*, 293–301.

CITATION

Han-Seung Lee, Xiao-Yong Wang, Li-Na Zhang and Kyung-Taek Koh, Analysis of the Optimum Usage of Slag for the Compressive Strength of Concrete, doi:10.3390/ma8031213

CHAPTER 3

Compressive Behavior of Fiber-Reinforced Concrete with End-Hooked Steel Fibers

Seong-Cheol Lee [1], Joung-Hwan Oh [2] and Jae-Yeol Cho [3]

[1]Department of NPP Engineering, KEPCO International Nuclear Graduate School, 658-91 Haemaji-ro, Seosaeng-myeon, Uljugun, Ulsan 689-882, Korea;
[2]Office of Offshore Wind Power, Korea Institute of Energy Technology Evaluation and Planning, Seoul 135-502, Korea;
[3]Department of Civil and Environmental Engineering, Seoul National University, Seoul 151-744, Korea

ABSTRACT

In this paper, the compressive behavior of fiber-reinforced concrete with end-hooked steel fibers has been investigated through a uniaxial compression test in which the variables were concrete compressive strength, fiber volumetric ratio, and fiber aspect ratio (length to diameter). In order to minimize the effect of specimen size on fiber distribution, 48 cylinder specimens 150 mm in diameter and 300 mm in height were prepared and then subjected to uniaxial compression. From the test results, it was shown that steel fiber-reinforced concrete (SFRC) specimens exhibited ductile behavior after reaching their compressive strength. It was also shown that the strain at the compressive strength generally increased along with an increase in the fiber volumetric ratio and fiber aspect ratio, while the elastic modulus decreased. With consideration for the effect of steel fibers, a model for the stress–strain relationship of SFRC under compression is proposed here. Simple formulae to predict the strain at the compressive strength and the elastic modulus of SFRC were developed as well. The proposed model and formulae will be useful for realistic predictions of the structural behavior of SFRC members or structures.

INTRODUCTION

Recently, as the demand for high strength concrete has increased, the structural behavior of reinforced concrete has become more brittle. In order to reduce this side effect, steel fiber-reinforced concrete (SFRC) has arisen as a viable method to attain ductility not only during post-cracking behavior under tension, but also during post-peak softening behavior under compression. For the last several decades, use of steel fibers has been limited to tunnel lining or crack control in concrete slabs as a non-structural material. Development of a rational compressive model for SFRC has been of importance since many researchers [1,2,3,4,5,6,7,8,9,10] have continuously conducted experimental and analytical research to exploit SFRC as a structural material.

Several researchers [11,12,13,14,15,16,17] have investigated the effect of steel fibers on the compressive behavior of SFRC for design purposes. Table 1 summarizes models of the compressive behavior of SFRC, which several independent research groups have developed based on their own test results. As shown in the table, in order to predict the compressive stress-strain response of SFRC, the measured compressive strength of SFRC is required in some models [11,15], while in the other models [12,13,14], the compressive strength of SFRC is derived from the compressive strength of plain concrete without fibers. In the models summarized in the table, the coefficient β, introduced first by Carreria and Chu [18] for normal concrete without fibers, has been modified to reflect the effect of steel fibers on compressive behavior. However, since the models proposed by Ezeldin and Balaguru [11], Someh and Saeki [13], and Nataraja et al. [15] were developed based on test results with crimped steel fibers, the appropriateness of the models has become questionable for concrete with end-hooked steel fibers.

The model presented by Hsu and Hsu [12] considers only several specific fiber volumetric ratios (0.50%, 0.75% and 1.00%), so its practical application is quite limited. In addition, in the experiments conducted by some research groups [12,13,14], only one fiber aspect ratio (length to diameter) was considered, so it is questionable whether the models proposed by those research groups reasonably represent the effect of the aspect ratio of steel fibers on compressive behavior. Moreover, some specimens [11,12] tested in the development of previous models were relatively small in comparison with fiber length, so the observed compressive behavior of SFRC could differ from that in real structures that are relatively large because fiber distribution can be significantly affected by the boundary surfaces of small specimens [19,20]. Some models [11,15] require data on the compressive strength of plain concrete without fibers in the same mix proportion as SFRC, but it is more practical to use the actual compressive strength of SFRC in the model. Consequently, a more reasonable model to represent the compressive behavior of SFRC should be developed from a more extensive experimental program using large specimens and various steel fiber configurations.

Table 1. Models for the compressive behavior of fiber reinforced concrete.

Researchers	Models
[1]Ezeldin and Balaguru [11]	$$f_c = f_c' \frac{\beta\left(\frac{\varepsilon}{\varepsilon_0}\right)}{\beta - 1 + \left(\frac{\varepsilon}{\varepsilon_0}\right)^\beta}$$ where, $f_c' = f_{cp}' + 11.232RI$; $\beta = 1.093 + 0.2429RI^{-0.926}$; $E_c = E_{cp} + 9936RI$; $\varepsilon_0 = \varepsilon_{0p} + 1427 \times 10^{-6}RI$; f_{cp}', ε_{0p}, and E_{cp} are the compressive strength, corresponding strain, and elastic modulus of plain concrete, respectively.
Hsu and Hsu [12]	$$f_c = f_c' \frac{n\beta\left(\frac{\varepsilon}{\varepsilon_0}\right)}{n\beta - 1 + \left(\frac{\varepsilon}{\varepsilon_0}\right)^{n\beta}} \text{ where for } 0 \leq \frac{\varepsilon}{\varepsilon_0} \leq \frac{\varepsilon_d}{\varepsilon_0}$$ $$f_c = 0.6f_c' \exp\left[-0.7\left(\frac{\varepsilon}{\varepsilon_0} - \frac{\varepsilon_d}{\varepsilon_0}\right)^{0.8}\right] \text{ for } \frac{\varepsilon_d}{\varepsilon_0} \leq \frac{\varepsilon}{\varepsilon_0}$$ where, ε_d is the strain at $0.6f_c'$ in the descending; f_c' is the measured compressive strength; $$\beta = \left(\frac{f_c'}{11.838(100V_f)^3 + 58.612}\right)^3 - 26V_f + 2.742; E_i = a_2 f_c' + C_2; \varepsilon_0 = a_1 f_c' + C_1$$ where, a_1, a_2, C_1, and C_2 are constant.
[1]Someh and Saeki [13]	$$f_c = f_c' \frac{\beta\left(\frac{\varepsilon}{\varepsilon_0}\right)}{\beta - 1 + \left(\frac{\varepsilon}{\varepsilon_0}\right)^\beta}$$ where, f_c' is the measured compressive strength; $\beta = 1.032\left[f_c'\left(1 + RI\right)\right]^{0.113}$; and $\varepsilon_0 = 0.00184f_c^{0.147}$.
[1]Mansur et al. [14]	$$f_c = f_c' \frac{\beta\left(\frac{\varepsilon}{\varepsilon_0}\right)}{\beta - 1 + \left(\frac{\varepsilon}{\varepsilon_0}\right)^\beta} \text{ for } 0 < \frac{\varepsilon}{\varepsilon_0} \leq 1$$ $$f_c = f_c' \frac{k_1\beta\left(\frac{\varepsilon}{\varepsilon_0}\right)}{k_1\beta - 1 + \left(\frac{\varepsilon}{\varepsilon_0}\right)^{k_2\beta}} \text{ for } 1 \leq \frac{\varepsilon}{\varepsilon_0}$$ where, f_c' is the measured compressive strength; $\beta = 1/\left[1 - \left(f_c'/\varepsilon_0 E_i\right)\right]$; $E_i = (10300 - 40000V_f) f_c'^{1/3}$; $\varepsilon_0 = (0.00050 + 0.000072RI) f_c'^{0.35}$.
[1]Nataraja et al. [15]	$$f_c = f_c' \frac{\beta\left(\frac{\varepsilon}{\varepsilon_0}\right)}{\beta - 1 + \left(\frac{\varepsilon}{\varepsilon_0}\right)^\beta}$$ where, $f_c' = f_{cp}' + 6.9133RI$; $\beta = 0.5811 + 0.8155RI^{-0.7406}$; $\varepsilon_0 = \varepsilon_{0p} + 0.00192RI$; f_{cp}' and ε_{0p} are the compressive strength of plain concrete and corresponding strain, respectively.

Note: [1]$RI = V_f l_f / d_f$ where V_f is a fiber volumetric ratio.

TEST PROGRAM FOR THE COMPRESSIVE BEHAVIOR OF SFRC

In this study, the compression test with ϕ 150 mm by 300 mm cylindrical specimens was conducted to investigate the compressive behavior of SFRC. The variables were concrete compressive strength (f'c), fiber volumetric ratio (Vf), and fiber aspect ratio (lf/df). Four fiber volumetric ratios ranging from 0.5% to 2.0% were considered, and end-hooked steel fibers with three different aspect ratios (47.6%, 63.6%, and 78.9%) were used. In order to investigate the effect of concrete compressive strength, two compressive strengths of 50 and 80 MPa were targeted for N and H series, respectively. In the test for H series, only steel fibers 30 mm in length and 0.38 mm in diameter were used since it was inferred that those fibers would be the most effective for optimizing ductile performance during post-peak compressive behavior, as shown in a previous test of the tensile behavior of SFRC [21]. Specifications of the steel fibers are presented in Table 2, and the notation for the test variables is illustrated in Figure 1.

Table 2. Properties of steel fibers.

Type	Length l_f [mm]	Diameter d_f [mm]	Tensile strength σ_{fu} [MPa]	Aspect ratio l_f/d_f
F1: RL-4550-BN *	50	1.05	1000	47.6
F2: RC-6535-BN *	35	0.55	1100	63.6
F3: RC-8030-BP *	30	0.38	2300	78.9

Note: * Fiber model names are given by the manufacturer, Bekaert.

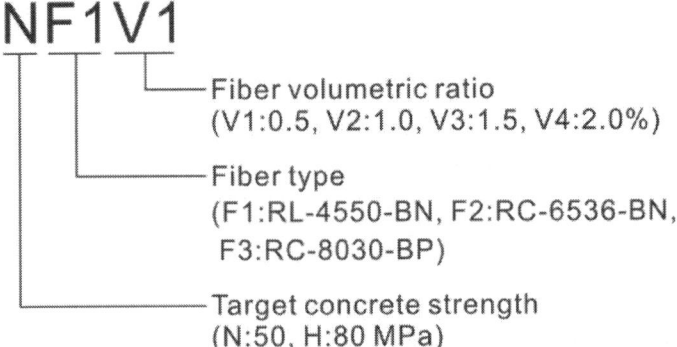

Figure 1. Test variables.

The concrete mix proportion of normal concrete without fibers is difficult to apply for SFRC since steel fibers reduce the workability for a given mixture design. Therefore, finer aggregate and cement should replace a portion of the coarse aggregate to guarantee both workability and target compressive strength. In this study, the concrete mixture design presented in Susetyo [22] was modified as presented in Table 3. The maximum aggregate size was 13 mm and Type I cement was used. Before mixing the concrete, the coarse aggregate was dried for one day after being submerged in water so that surface saturation could be achieved when the concrete was mixed. The mixing procedure presented in Susetyo [22] was followed, and cylindrical specimens were cast in two phases. After each half casting, the specimens were vibrated for 5 seconds on a vibrating table at 5 Hz.

Table 3. Mix proportions.

Concrete strength	Water to binder ratio	Water [kg/m³]	Cement [kg/m³]	Silica fume [kg/m³]	Sand [kg/m³]	Coarse aggregate [kg/m³]	Super-plasticizer [kg/m³]
Normal strength (N)	0.35	200	572	-	798	627	1.430
High strength (H)	0.25	200	737	64	667	569	6.008

Three cylinders per test series were cast, which resulted in 48 specimens for 16 test series. The specimens were demolded one day after casting and then cured at 20 °C and 70% relative humidity until the age of 28 days, when the compression tests were conducted. To prevent the specimens from being subjected to possible eccentric loading, both the top and bottom surfaces of all specimens were ground. Two linear variable differential transformers (LVDTs) with a reference length of 150 mm were mounted on the circumferential side of the specimens so that the average longitudinal strain could be simultaneously measured during the compression test. The specimens were subjected to uniaxial compression under a stroke control of 0.4 mm/min.

TEST RESULTS AND INVESTIGATIONS

Pre-Peak Compressive Behavior

The pre-peak compressive behavior of SFRC can be represented by several characteristics such as concrete compressive strength, compressive strain at the peak, and elastic modulus. These characteristics, as measured from the compression test, are summarized in Table 4. The test results were obtained from the average for three specimens per test series.

The effect of the fiber volumetric ratio on compressive strength can be seen in Figure 2. As shown in the figure, the compressive strength increased slightly with increasing fiber volumetric ratio in the NF1 series, while it decreased in the NF3 series. This inconsistency has been noted by other researchers as well. Ezeldin and Balaguru [11] reported that the compressive strength increased with increasing steel fiber volume due to the transverse confinement effect of the steel fibers, which restrained the lateral expansion of SFRC specimens. On the other hand, Hsu and Hsu [12] reported that steel fibers did not contribute to an increase in the compressive strength since more voids could be produced in SFRC because of its low workability. These effects of steel fibers can also be explained with the measured slumps presented in Table 4. A large slump with good workability was observed in the NF1 series, in which relatively long steel fibers with a small aspect ratio were mixed, resulting in fewer steel fibers per unit of concrete volume. On the other hand, when NF3 series with shorter steel fibers with a large aspect ratio were mixed, resulting in a large number of steel fibers per unit concrete volume, a small slump was observed. It can be concluded, therefore, that the compressive strength of SFRC can be affected by the fiber aspect ratio as well as the fiber volumetric ratio.

Table 4. Test results.

Specimen.	Target strength [MPa]	l_f/d_f	V_f [%]	f_c' [MPa]	ε_0 [$\mu\varepsilon$]	E_c [MPa]	Slump [mm]
NF1V1			0.5	48.7	3,137	25,406	146
NF1V2		45	1.0	49.0	3,047	25,781	154
NF1V3			1.5	51.1	3,190	25,187	129
NF1V4			2.0	51.2	3,244	26,091	111
NF2V1			0.5	45.9	2,701	29,987	113
NF2V2	50	65	1.0	43.0	3,150	22,812	135
NF2V3			1.5	34.5	2,289	26,107	141
NF2V4			2.0	41.4	3,187	21,758	83
NF3V1			0.5	49.8	3,078	24,373	111
NF3V2		80	1.0	43.8	3,100	26,661	114
NF3V3			1.5	44.4	3,422	20,640	83
NF3V4			2.0	36.7	3,499	22,880	39
HF3V1			0.5	81.9	3,191	31,328	225
HF3V2	80	80	1.0	85.7	3,424	33,581	193
HF3V3			1.5	82.8	3,288	33,855	164
HF3V4			2.0	83.0	3,922	28,246	118

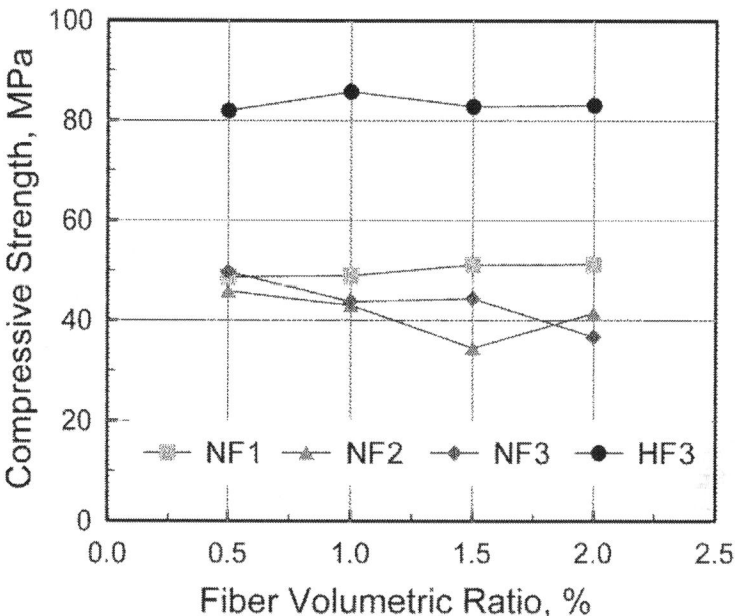

Figure 2. Measured compressive strength.

The effect of the concrete mix proportion was also investigated by comparing NF3 and HF3 series, which contain the same fibers. As seen in Table 4 and Figure 2, in the NF3 series, the compressive strength generally decreased with an increase in the fiber volumetric ratio. In the HF3 series, on the other hand, the effect of the fiber volumetric ratio on the compressive strength was not obvious since good workability with a large slump was obtained by adopting a different mixture design, *i.e.*, silica fume was added to achieve higher compressive strength in the H series. Aside from NF2V3 and NF3V4 which exhibited the compressive strengths scattered from the others, it can be seen that the compressive strength of SFRC is not significantly affected by the fiber volumetric ratio.

Figure 3 shows the effect of the fiber volumetric ratio on the strain at the peak stress. As presented in the figure, in NF3 and HF3 series, in which shorter steel fibers with a high aspect ratio were mixed, the strain corresponding to the compressive strength significantly increased with an increase in the fiber volumetric ratio. This tendency was not clearly observed in the NF1 series, in which long steel fibers with a low aspect ratio were mixed. On the other hand, the other members except NF2V3 exhibited the general tendency regarding the effect of the fiber volumetric ratio. The test results indicate that the strain at the compressive strength is affected by both the fiber volumetric ratio and fiber aspect ratio, which correspond to the number of fibers per unit concrete volume.

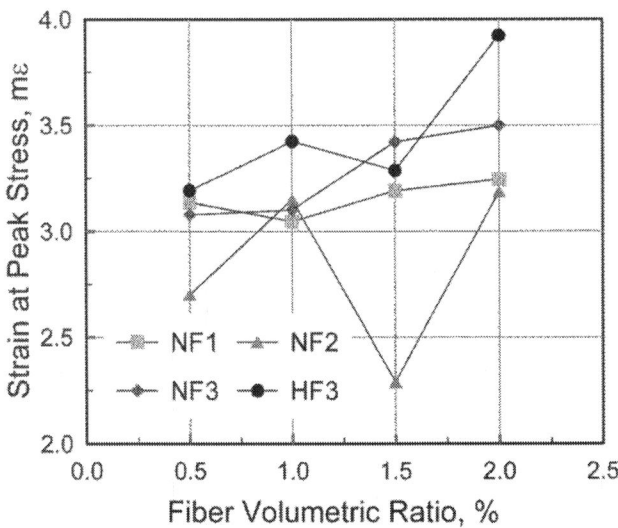

Figure 3. Measured compressive strain at the compressive strength.

From the measured compressive stress-strain responses, the elastic modulus was evaluated according to ASTM C469 [23] as follows:

$$E_c = \frac{f_{c2} - f_{c1}}{\varepsilon_{c2} - \varepsilon_{c1}} \tag{1}$$

where f_{c1} and f_{c2} are the compressive stresses corresponding to $\varepsilon_{c1} = 50 \times 10^{-6}$ and ε_{c2} at $0.4f'c$, respectively; and $f'c$ is the measured compressive strength.

The evaluated elastic modulus is presented in Figure 4. As shown in the figure, the elastic modulus in the NF2 and NF3 series generally decreased with an increase in the fiber volumetric ratio, while the elastic modulus in the NF1 series remained almost constant regardless of the fiber volumetric ratio. In the HF3 series, however, since fewer voids existed in the specimens because of the relatively high slump due to the different mixture design, it was not obvious that the fiber volumetric ratio had a significant effect on the elastic modulus when the fiber volumetric ratio was not greater than 1.5%.

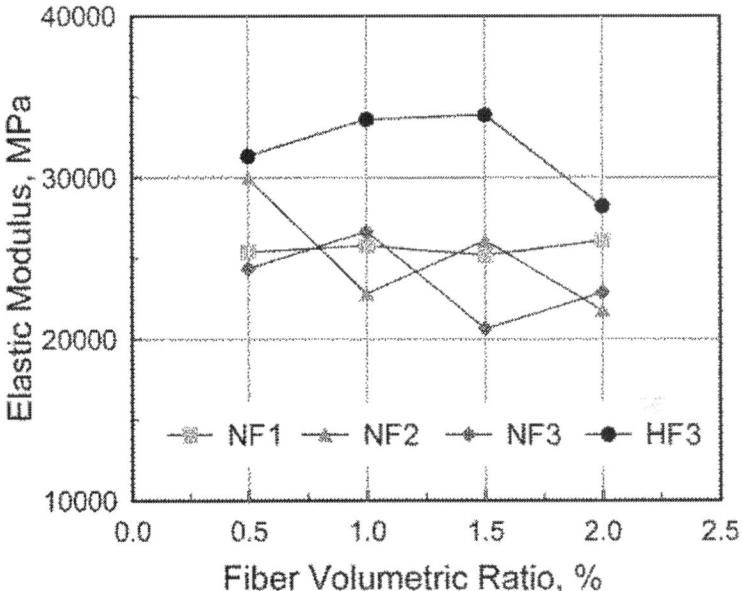

Figure 4. Measured elastic modulus.

Post-Peak Compressive Behavior

While normal concrete without steel fibers shows a drastic decrease in stress after experiencing compressive strength, it is well known that SFRC can exhibit ductile behavior even after reaching its compressive strength because of the transverse confinement effect of the steel fibers [11,12,13,14,15]. This ductile behavior was also indicated by the test results in this study, as shown in Figure 5 and Figure 6. In these figures, the measured compressive stresses and strains are normalized by the compressive strength and the strain at the compressive strength, respectively, because the measured compressive strength and the corresponding strain varied depending on the fiber volumetric ratio and fiber aspect ratio.

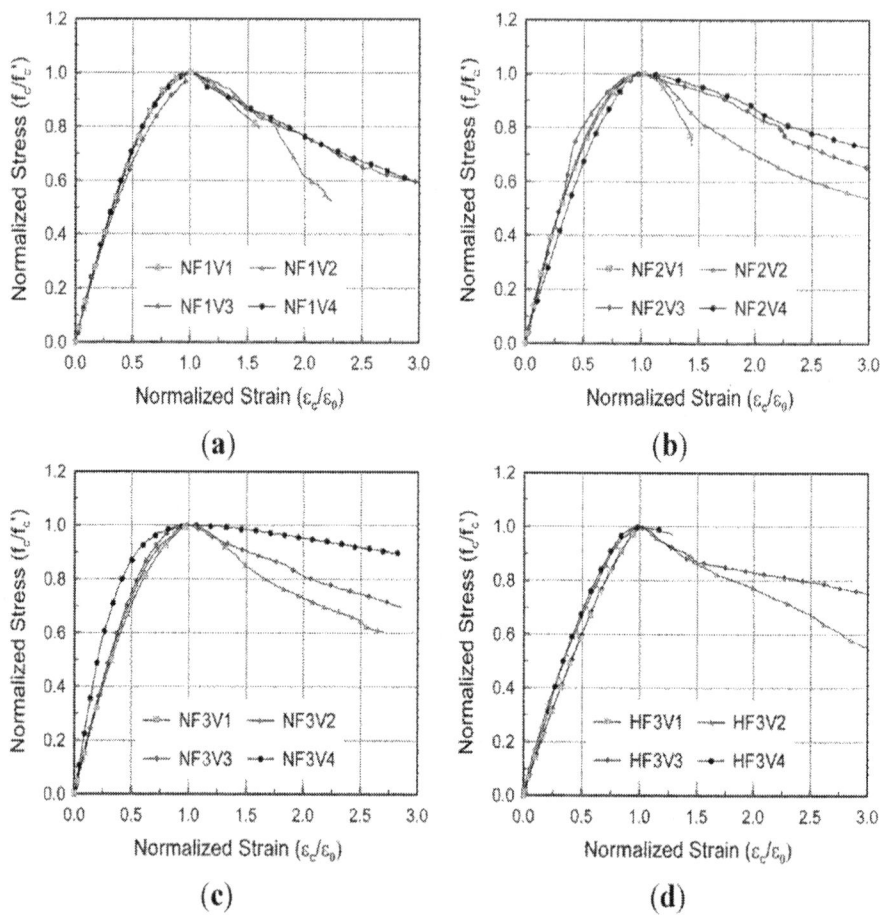

Figure 5. Effect of fiber volumetric ratio on the stress-strain response under compression. (**a**) NF1 series; (**b**) NF2 series; (**c**) NF3 series; (**d**) HF3 series.

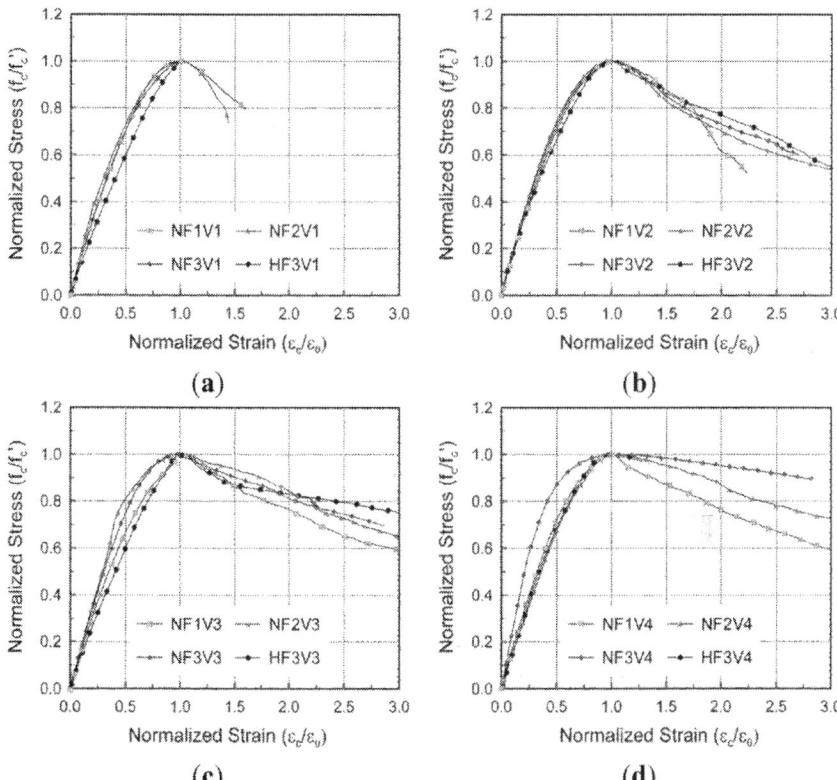

Figure 6. Effect of fiber aspect ratio on the stress-strain response under compression. (**a**) NF1 series; (**b**) NF2 series; (**c**) NF3 series; (**d**) HF3 series.

In order to investigate the effect of test variables on post-peak compressive behavior, one of the three test results extensively measured during large compressive strains was chosen for each test series. The post-peak compressive behavior for some specimens with a relatively low fiber volumetric ratio (V1 series) or with high compressive strength of concrete (H series) could not be measured until the ultimate fracture because the applied load drastically dropped upon failure. As shown in Figure 5, more residual compressive stress after peak stress was observed in the specimens with a larger fiber volumetric ratio. As shown in Figure 6, the specimens containing fibers with a higher aspect ratio showed more ductile post-peak compressive behavior, since a greater transverse confinement effect could be attained by the better pull-out behavior of the steel fibers. This tendency, an effect of the fiber aspect ratio, was investigated through tensile tests, in which the NF3 series generally showed the most desirable performance in terms of tensile behavior [21].

MODEL FOR THE COMPRESSIVE BEHAVIOR OF SFRC MEMBERS

As discussed in the previous section, the compressive behavior of SFRC is significantly affected by the fiber volumetric ratio, fiber aspect ratio, and compressive strength. In this section, the simple formulae for the elastic modulus and the strain at the compressive strength, which represents the pre-peak compressive behavior of SFRC, are derived from the test results. Finally, a model for the compressive behavior of SFRC including the post-peak behavior will be presented with consideration for the effect of steel fibers by employing the fiber reinforcing index, $RI = V_f l_f / d_f$, which was introduced by previous researchers [11,13,14,15] (Table 1). The models proposed in this paper will be verified through comparison with the test results. The models presented in Table 1 will also be compared with the proposed model as well as the test results.

Strain at the Compressive Strength
Based on the effect of steel fibers for which the strain at the compressive strength is significantly affected by both the fiber volumetric ratio and fiber aspect ratio, the following equation to predict the strain at the compressive strength of SFRC was derived from regression analysis to minimize the square root error corresponding to differences between predictions and test results.

$$\varepsilon_0 = \left(0.0003 V_f \frac{l_f}{d_f} + 0.0018 \right) f_c'^{0.12} = (0.0003 RI + 0.0018) f_c'^{0.12}$$

$$(2)$$

where f'_c is the compressive strength of SFRC. In Figure 7, the strains at the compressive strength measured from the compression test are compared with predictions made using Equation (2) and by other researchers [12,13,14]. Although some researchers [11,15] also proposed their own simple equations to predict the strain at the compressive strength, their models are not included in this comparison because those models are based on the compressive strength of plain concrete. The test results and the model presented in Equation (2) show that the strain at the peak stress gradually increases with increasing fiber reinforcing index (Figure 7). As shown in the figure, Hsu and Hsu [12] predicted the strain at the compressive strength with an acceptable margin, but its application is limited to the specific fiber content outside the common practical range of 1.0%–1.5%, as previously mentioned. Someh and Saeki [13] predicted that the strain at the compressive strength would be nearly constant regardless of the steel fiber content because only the compressive strength of SFRC was taken into account. Mansur et al. [14] predicted the increasing tendency of strain at peak stress with an increase in the fiber reinforcing index, but they significantly underestimated the strain at the compressive strength. This result and investigation apply to both N and H series. It can be concluded, therefore, that the strain at the compressive strength can be more reasonably predicted by the model proposed in this paper.

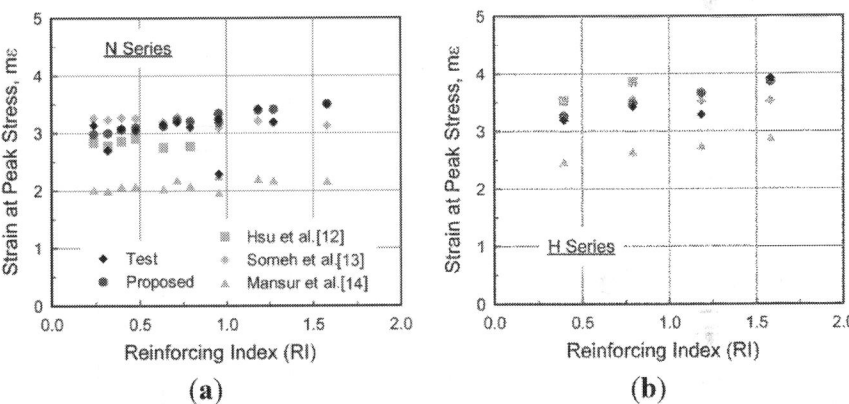

Figure 7. Comparison of the strain at the compressive strength. (**a**) N Series; (**b**) H Series.

Elastic Modulus

In the same manner as the strain at the compressive strength, the following equation for the elastic modulus was derived from regression analysis:

$$E_c = \left(-367V_f \frac{l_f}{d_f} + 5520\right) f_c'^{0.41} \tag{3}$$

In Figure 8, the elastic modulus predicted by the equation was compared with the test results and predictions of previous researchers [12,14]. The model proposed by Ezeldin and Balaguru [11] is not included in the comparison because the compressive strength of plain concrete without steel fibers using the same mixture as SFRC is required to predict the elastic modulus. As presented in the figure, Mansur *et al.* [14] predicted how the elastic modulus would vary, but they generally overestimated the elastic modulus, since the elastic modulus was evaluated from the tangential stiffness at initial loading. This differs from the evaluation procedure used in this paper, in which the secant stiffness between the initial and 40% of the compressive strength was used for the elastic modulus, as presented in ASTM C469 [23]. Although the predictions made by Hsu and Hsu [12] were consistent with the test results for N series, the application of their model is limited to specific fiber levels. On the other hand, extensive testing revealed that Equation (3) accurately predicted the elastic modulus by considering the effect of steel fibers.

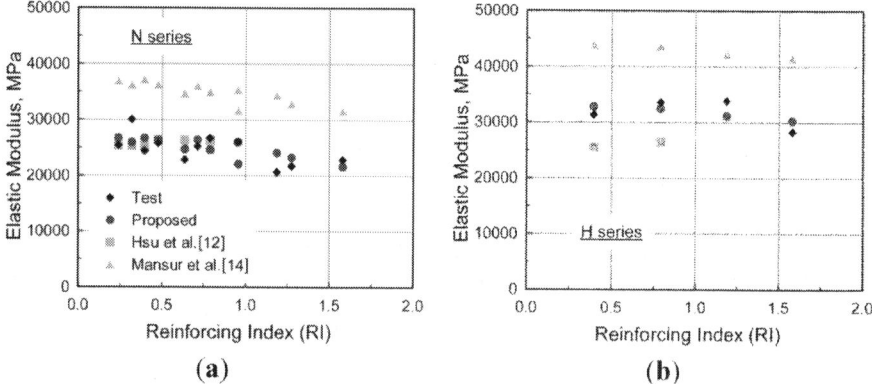

(a)   (b)

Figure 8. Comparison of the elastic modulus. (**a**) N Series; (**b**) H Series.

Compressive Stress-Strain Relationship

To simulate the stress-strain response of SFRC under compression, the two phases of pre- and post-peak behaviors should be considered separately because no significant lateral crack opening is observed in pre-peak compressive behavior, while the transverse confinement effect of steel fibers to restrain lateral crack opening is considerable in post-peak compressive behavior. As previous research groups' models summarized in Table 1 are based on the model presented by Carreira and Chu [18], the following equation was employed to represent the compressive behavior of SFRC:

$$f_c = f_c' \left[\frac{A(\varepsilon_c/\varepsilon_0)}{A - 1 + (\varepsilon_c/\varepsilon_0)^B} \right] \tag{4}$$

where A and B are parameters used to reflect the effect of steel fibers on compressive behavior. For the pre-peak compressive behavior, parameters A and B in Equation (4) can be calculated by adopting the basic form presented by Carreira and Chu [18] and Mansur et al.[14] as follows:

$$A = B = \frac{1}{1 - \left(\dfrac{f_c'}{\varepsilon_0 E_c} \right)} \quad \text{for } \varepsilon_c/\varepsilon_0 \leq 1.0 \tag{5}$$

where ε_0 and E_c can be calculated from Equations (2) and (3), respectively.

For the post-peak compressive behavior, parameters A and B have been derived through regression analysis by employing the fiber reinforcing index, Vflf/df.

$$A = 1 + 0.723 \left(V_f \frac{l_f}{d_f} \right)^{0.957} \quad \text{for } \varepsilon_c/\varepsilon_0 > 1.0 \tag{6}$$

$$B = \left(\frac{f_c'}{50} \right)^{0.064} \left[1 + 0.882 \left(V_f \frac{l_f}{d_f} \right)^{-0.882} \right] \tag{7}$$

$$> A \text{ in Equation (6) for } \varepsilon_c/\varepsilon_0 > 1.0$$

B calculated by Equation (7) should not be less than A calculated by Equation (6) to prevent the compressive stress in the post-peak behavior from being higher than the compressive strength. This limitation on the parameters can be determined through the derivative of Equation (4) for the post-peak compressive behavior.

Figure 9 compares the test results and predictions made by the model proposed in this paper and by other researchers [11,12,13,14,15]. The concrete compressive strengths and corresponding strains of SFRC measured through the tests were used in the model predictions presented by Ezeldin and Balaguru [11] and Nataraja *et al.* [15], who proposed that the compressive strength and the corresponding strains of SFRC be calculated from the compressive strength of specimens without steel fibers under the same mixture design as SFRC. As presented in the figures, although the measured compressive strength and corresponding strain were used, both Ezeldin and Balaguru [11] and Nataraja *et al.* [15] significantly overestimated the residual stress in the post-peak compressive behavior. Moreover, a stiffer compressive response was predicted before the compressive strength was reached. This indicates that the effect of steel fibers on the compressive behavior cannot reasonably be taken into an account in those models, even though the actual compressive strength and corresponding strain are used. This result might due to specimens being small compared to the fiber lengths [11], or a different type of fiber, *i.e.,* crimped fiber being used in their test programs [15].

Hsu and Hsu [12] generally made good predictions of the compressive behavior of the specimens with a low fiber volumetric ratio and fibers with a low aspect ratio (see NF1V1 and NF1V2), but they underestimated the residual stress on the post-peak compressive behavior of the specimens with steel fibers for which the aspect ratio was relatively high (see NF3V1 and NF3V2). In other words, since the model presented by Hsu and Hsu [12] only considers the effect of the fiber volumetric ratio, the effect of the fiber aspect ratio cannot be considered in the prediction of the post-peak compressive behavior of SFRC. Moreover, this model is only applicable to specimens in which the fiber volumetric ratio is not greater than 1.0%.

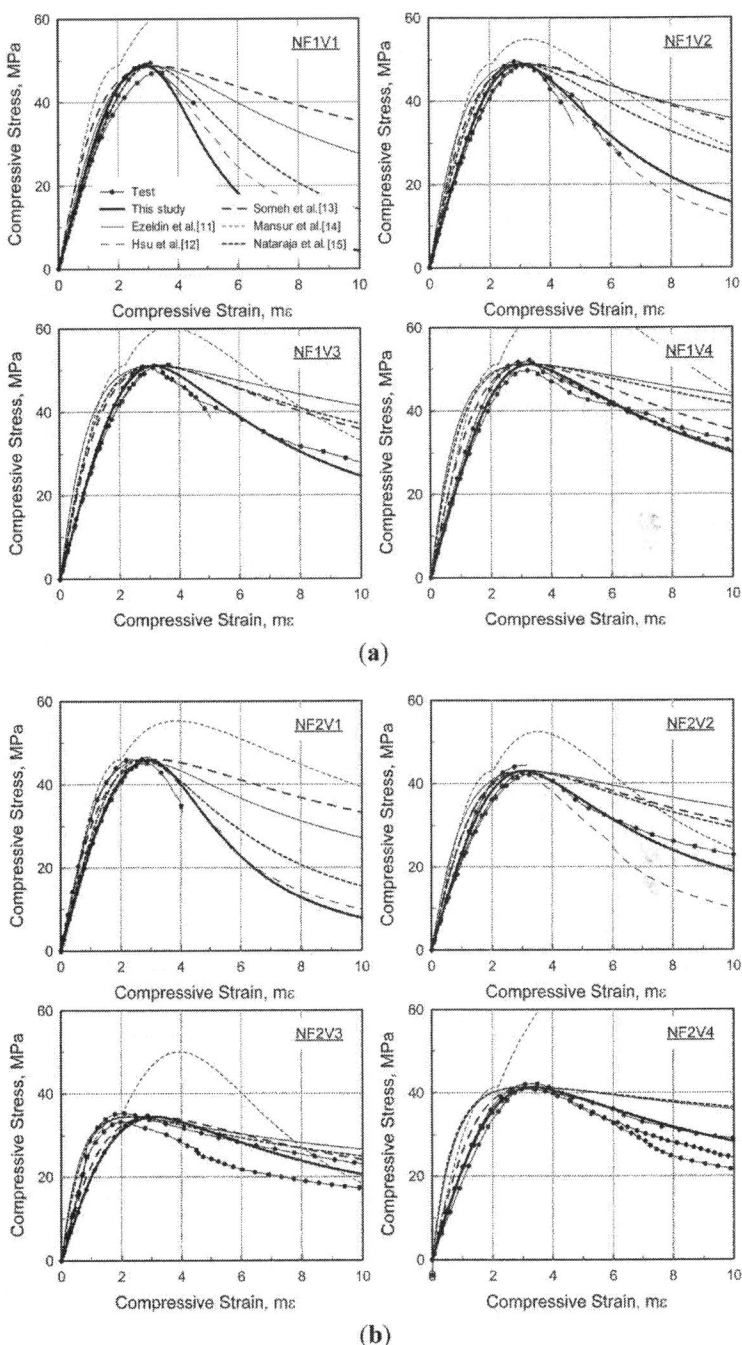

Figure 9. Comparison of the stress-strain response under compression. (**a**) NF1 series; (**b**) NF2 series; (**c**) NF3 series; (**d**) HF3 series.

Someh and Saeki [13] successfully predicted only the compressive behavior of specimens with a high fiber volumetric ratio and steel fibers with a high aspect ratio (see NF3V3, NF3V4 and HF3V3). For other specimens, the residual stress in the post-peak compressive behavior was significantly overestimated. The prediction of Mansur et al. [14] was not applicable since the compressive stress in the post-peak behavior was higher than the compressive strength when k_1 was larger than k_2 in the model summarized in Table 1. On the other hand, the compressive behavior predicted by Equations (4)–(7) proposed in this paper showed good agreement with the test results for various concrete compressive strengths and fiber contents.

CONCLUSIONS

Although many researchers proposed several models from their own test results in order to quantitatively evaluate the compressive behavior of SFRC, the appropriateness of the previous models is still questionable since the variables were too limited to extensively reflect the effect of steel fibers, or since the test specimens for the investigation were relatively small compared to the fiber length. The compressive behavior may have been different from those in real structures because of different fiber distributions.

In this paper, an experimental program was implemented to investigate the compressive behavior of SFRC with end-hooked steel fibers. The variables of the test program were concrete compressive strength, fiber volumetric ratio, and fiber aspect ratio. To minimize the effect of member size on fiber distribution, large cylindrical specimens were prepared and tested. Comparison between the predictions of other researchers' models and the test results revealed that no previous model could acceptably reflect the effect of end-hooked steel fibers on the compressive behavior.

In order to better represent the pre-peak compressive behavior of SFRC, simple formulae to predict the strain at the compressive strength and the elastic modulus were derived from the test results. In addition, by modifying the coefficient β, a simple model to predict the compressive behavior of SFRC was proposed, with consideration for the effect of steel fibers. The compressive behavior of SFRC with end-hooked steel fibers can be reasonably predicted using the proposed formulae and model. The proposed model may prove useful for predicting the structural behavior of SFRC members or structures with end-hooked steel fibers. Further study is required to understand the effect of other fiber types, such as straight steel fibers and synthetic fibers. In addition, through more investigation for the effect of SFRC member size on the compressive behavior, the proposed model can be applied to SFRC members with various dimensions.

ACKNOWLEDGMENTS

This research was supported by Integrated Research Institute of Construction and Environmental Engineering at Seoul National University. The authors wish to express their gratitude for the support.

AUTHOR CONTRIBUTIONS

Seong-Cheol Lee and Jae-Yeol Cho managed the research in this paper and also prepared the manuscript together; Joung-Hwan Oh conducted the experimental program and derived the empirical models.

REFERENCES

1. Ashour, S.A.; Hasanain, G.S.; Wafa, F.F. Shear behavior of high-strength fiber reinforced concrete beams. *ACI Struct. J.* 1992, *89*, 176–184.
2. Casanova, P.; Rossi, P.; Schaller, I. Can steel fibers replace transverse reinforcement in reinforced concrete beams? *ACI Mater. J.* 1997, *94*, 341–354.
3. Noghabai, K. Beams of fibrous concrete in shear and bending: Experiment and model. *ASCE J. Struct. Eng.* 2000, *126*, 243–251.
4. Meda, A.; Plizzari, G.A. New design approach for steel fiber-reinforced concrete slabs-on-ground based on fracture mechanics. *ACI Struct. J.* 2004, *101*, 298–303.

5. Minelli, F.; Vechcio, F.J. Compression field modeling of fiber-reinforced concrete members under shear loading. *ACI Struct. J.* 2006, *103*, 244–252.

6. Dinh, H.H.; Parra-Montesinos, G.J.; Wight, J.K. Shear behavior of steel fiber-reinforced concrete beams without stirrup reinforcement. *ACI Struct. J.* 2010, *107*, 597–606.

7. Susetyo, J.; Gauvreau, P.; Vecchio, F.J. Effectiveness of steel fiber as minimum shear reinforcement. *ACI Struct. J.* 2011, *108*, 488–496.

8. Hwang, J.-H.; Lee, D.H.; Ju, H.; Kim, K.S.; Seo, S.-Y.; Kang, J.-W. Shear behavior models of steel fiber reinforced concrete beams modifying softened truss model approaches. *Materials* 2013, *6*, 4847–4867.

9. Deluce, J.R.; Lee, S.-C.; Vecchio, F.J. Crack model for steel fiber-reinforced concrete members containing conventional reinforcement. *ACI Struct. J.* 2014, *111*, 93–102.

10. Shahnewaz, M.; Alam, M.S. Improved shear equations for steel fiber-reinforced concrete deep and slender beams. *ACI Struct. J.* 2014, *111*, 851–860.

11. Ezeldin, A.S.; Balaguru, P.N. Normal and high strength fiber reinforced concrete under compression. *ASCE J. Mater. Civil Eng.* 1992, *4*, 415–429.

12. Hsu, L.S.; Hsu, C.T.T. Stress-strain behavior of steel-fiber high-strength concrete under compression. *ACI Struct. J.* 1994, *91*, 448–457.

13. Someh, A.K.; Saeki, N. Prediction for the stress-strain curve of steel fiber reinforced concrete. *Proc. Jpn. Concr. Inst.* 1994, *18*, 1149–1154.

14. Mansur, M.A.; Chin, M.S.; Wee, T.H. Stress-strain relationship of high-strength fiber concrete in compression. *ASCE J. Mater. Civil Eng.* 1999, *11*, 21–29.

15. Nataraja, M.; Dhang, N.; Gupta, A. Stress-strain curves for steel-fiber reinforced concrete under compression. *Cement Concr. Compos.* 1999, *21*, 383–390.

16. Bencardino, F.; Rizzuti, L.; Spadea, G.; Swamy, R.N. Stress-strain behavior of steel fiber-reinforced concrete in compression. *J. Mater. Civil Eng.* 2008, *20*, 255–263.

17. Rizzuti, L.; Bencardino, F. Effects of fibre volume fraction on the compressive and flexural experimental behavior of SFRC. *Contemp. Eng. Sci.* 2014, *7*, 379–390.

18. Carreira, D.J.; Chu, K.H. Stress-strain relationship for plain concrete in compression. *ACI J.* 1985, *82*, 797–804.

19. Lee, S.-C.; Cho, J.-Y.; Vecchio, F.J. Diverse embedment model for steel fiber-reinforced concrete in tension: Model development. *ACI Mater. J.* 2011, *107*, 516–525.

20. Lee, S.-C.; Cho, J.-Y.; Vecchio, F.J. Diverse embedment model for steel fiber-reinforced concrete in tension: Model verification. *ACI Mater. J.* 2011, *107*, 526–535.
21. Oh, J.-H. Uniaxial Behavior of Steel Fiber Reinforced Concrete. Master's Thesis, Seoul National University, Seoul, Korea, 28 August 2011.
22. Susetyo, J. Fibre Reinforcement for Shrinkage Crack Control in Prestressed, Precast Segmental Bridges. Ph.D. Thesis, University of Toronto, Toronto, ON, Canada, 2009.
23. *ASTM C 469 Standard Test Method for Static Modulus of Elasticity and Poisson's Ratio of Concrete in Compression*; Annual Book of ASTM Standards. American Society for Testing and Materials: West Conshohocken, PA, USA, 2002.

CITATION

Seong-Cheol Lee, Joung-Hwan Oh and Jae-Yeol Cho, Compressive Behavior of Fiber-Reinforced Concrete with End-Hooked Steel Fibers, doi:10.3390/ma8041442

CHAPTER 4

Compressive Behavior of Concrete Confined with GFRP Tubes and Steel Spirals

Liang Huang [1], Xiaoxun Sun [1], Libo Yan [2,3] and Deju Zhu [1]

[1]College of Civil Engineering, Hunan University, Changsha 410082, China;
[2]Department of Organic and Wood-Based Construction Materials, Technical University of Braunschweig, Hopfengarten 20, Braunschweig 38102, Germany;
[3]Centre for Light and Environmentally-Friendly Structures, Fraunhofer Wilhelm-Klauditz-Institut WKI, Bienroder Weg 54E, Braunschweig 38108, Germany

ABSTRACT

This paper presents the experimental results and analytical modeling of the axial compressive behavior of concrete cylinders confined by both glass fiber-reinforced polymer (GFRP) tube and inner steel spiral reinforcement (SR). The concrete structure is termed as GFRP–SR confined concrete. The number of GFRP layers (1, 2, and 3 layers) and volumetric ratios of SR (1.5% and 3%) were the experimental variables. Test results indicate that both GFRP tube and SR confinement remarkably increase the ultimate compressive strength, energy dissipation capacity, and ductility of concrete. The volumetric ratio of SR has a more pronounced influence on the energy dissipation capacity of confined concrete with more GFRP layers. In addition, a stress–strain model is presented to predict the axial compressive behavior of GFRP–SR confined concrete. Comparisons between the analytical results obtained using the proposed model and experimental results are also presented.

INTRODUCTION

In the last two decades, the use of fiber-reinforced polymer (FRP) composites has drawn much attention in civil engineering. Lateral confinement using FRP or spiral reinforcement (SR) has been demonstrated to increase compression strength, deformability, and energy absorption capacity of concrete [1,2]. Confinement of concrete with externally bonded FRP is an important application of FRP composites. Numerous experimental studies have been conducted to examine the performance of FRP composites in retrofitting existing concrete columns [3,4,5,6,7,8,9,10,11,12,13,14,15,16,17,18]. Recently, research efforts have been directed towards the applications of FRP in new column constructions; concrete-filled FRP tubes (CFFTs) have been used as high-performance composite columns in construction of earthquake-resistant structures [19,20,21,22,23,24,25,26,27,28,29]. These studies showed that the stress–strain curve of well-confined concrete with FRP is characterized by two ascending branches with increasing ultimate concrete compressive strength and strain. Moreover, the two commonly used FRP composites in FRP-confined cylindrical concrete specimens are carbon FRP (CFRP) and glass FRP (GFRP) composites, which can reach almost the same level of effectiveness, but at different axial strain levels, which renders more attractive the use of GFRP jackets that also exploit ductility while maintaining the same effectiveness of CFRP jackets [30].

The spirally reinforced column is an important practical example of the use of concrete under three-dimensional compression [31] and the post-peak behavior of concrete confined with SR has been demonstrated to be very ductile [1,31]. The final failure of FRP-confined concrete corresponding to the rupture of FRP is very sudden and explosive [3,4,5,6,7,8,9,10,11,12,13,14,15,16,17,18,19,20,21,22,23,24,25,26,27,28,29] because of the linear elastic tensile stress–strain behavior of FRP, thus, relatively high compression strength as well as high ductility are expected for concrete under combined FRP–SR confinement, *i.e.*, in FRP–SR confined concrete. Few tests have been performed to investigate the behavior of concrete confined with both transverse steel reinforcement (TSR) and FRP [32,33,34,35,36,37,38,39,40,41,42]. Moreover, most of these tests were conducted to examine the performance of FRP jackets in retrofitting existing reinforced concrete (RC) columns that contained small amounts of TSR, which did not influence the behavior of FRP confined concrete. The lack of information on the behavior of FRP–TSR-confined concrete, which was designed as a high-performance composite system used in construction of earthquake-resistant structures, was noted by a number of researchers, e.g., De Lorenzis and Tepfers [43], Teng and Lam [44], and Monti [30]. To understand the effect of TSR on the effectiveness of FRP, Monti [30], as well as other researchers, aimed to investigate the interaction between FRP and TSR. In the present paper, the longitudinal reinforcement component in the aforementioned studies was removed, because this component disturbs the analysis of 3-D compression of concrete and influences the "pure" transverse confinement circumstance provided by FRP and spiral reinforcement (SR). Currently, our research team is conducting a series of experimental studies on fundamental dynamic behaviors of GFRP–SR confined concrete under high strain rate compressive stresses by using a large-capacity drop-hammer machine. In addition to the importance of understanding fundamental mechanical behaviors of GFRP–SR confined concrete, the present study is a basic component for future studies.

Over the past two decades, a large number of constitutive models were developed for FRP confined concrete [41,42,43,44,45, 46,47,48,49,50,51,52,53,54,55,56]. However, most of these confinement models are suitable only for concrete confined in a single material, either internal TSR or outer FRP tube. Recently, several models have been proposed to describe the axial and lateral behavior of concrete confined by both TSR and FRP composites [35,57,58,59]. The suitability of existing confinement models for FRP wrapping confined RC column for GFRP–SR confined concrete, which was considered in the present study, is unknown. Thus, this study investigated the compressive behavior of normal strength concrete confined with both outer GFRP tube and internal SR. The effectiveness of existing confinement models for FRP wrapping confined reinforced concrete was also evaluated. Based on the experimental results, a design-oriented confinement stress–strain model was also developed.

EXPERIMENTAL SECTION

Test Matrix

A total of 18 GFRP–SR confined concrete cylinders with a diameter of 150 mm and a height of 300 mm were constructed and tested under axial compression. The number of GFRP layers (n_{FRP}) and SR volumetric ratio (ρ_s) were the main experimental parameters. Cylinder specimens were identified by two sets of characters as follows: the first set indicated the number of GFRP layers (*i.e.*, P1, P2, and P3 indicated one, two, and three layers, respectively); and the second set of characters indicated the pitch of SR (*i.e.*, S1 = 25 mm and S2 = 50 mm). Table 1 summarizes the testing matrix.

Table 1. Details of confined concrete cylinder specimens (150 mm × 300 mm).

Specimen	f_c' (MPa)	n_{FRP}	E_f (GPa)	t (mm)	f_{lf}/f_c'	ρ_f (%)	f_{hy} (MPa)	ρ_s (%)	f_{ls}/f_c'	Number of Specimens
P1S1	30	1	60.8	0.436	0.190	20.20	356	3.0	0.164	3
P2S1	30	2	60.8	0.872	0.378	25.23	356	3.0	0.164	3
P3S1	30	3	60.8	1.308	0.567	27.52	356	3.0	0.164	3
P1S2	30	1	60.8	0.436	0.190	20.20	356	1.5	0.073	3
P2S2	30	2	60.8	0.872	0.378	25.23	356	1.5	0.073	3
P3S2	30	3	60.8	1.308	0.567	27.52	356	1.5	0.073	3

The equivalent steel-confined concrete concept is displayed in Figure 1. The thickness, e, is given by:

$$e = K_e \frac{A_{shy}}{2s} \tag{1}$$

and the geometric effectiveness coefficient of SR, K_e [48], is expressed by:

$$K_e = \frac{(1 - 0.5\frac{S'}{D})}{(1 - \rho s)} \tag{2}$$

where s is the pitch of spirals; A_{shy} is the total cross-sectional area of the spirals in the longitudinal direction; D is the full cylinder diameter; S' is the clear pitch of spirals (edge to edge); and $\rho_s = 2A_{shy}/Ds$ is the volume ratio of SR to total volume of core, which is measured center-to-center of the spiral.

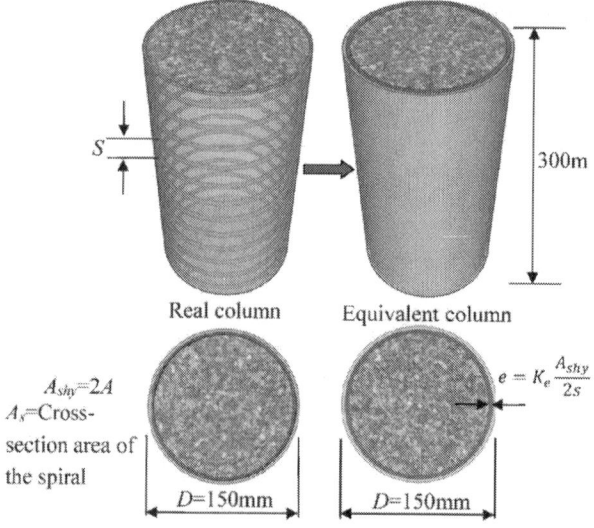

Real column   Equivalent column

$A_{shy} = 2A$
A_s=Cross-section area of the spiral

$e = K_e \frac{A_{shy}}{2s}$

D=150mm D=150mm

300m

Figure 1. Equivalent steel-confined concrete concept.

The effective pressure because of the action of the lateral steel, f_{ls}, is derived from force equilibrium at half cross section of the confined concrete cylinder [48], as illustrated in Figure 2:

$$f_{ls} = \frac{2ef_h}{D} = \frac{K_e \frac{A_{shy}}{s} f_h}{D} = \frac{1}{2}\rho_{sey} f_h \tag{3}$$

where f_h is the lateral steel stress and ρ_{sey} is the effective sectional ratio of the confining reinforcement, which is equal to $K_e\rho_s$.

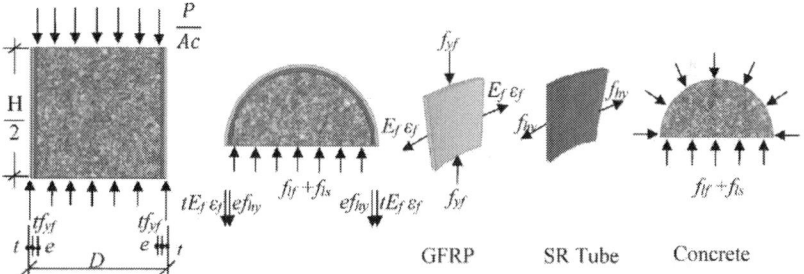

Figure 2. Confinement action of GFRP-SR-confined concrete.

The lateral pressure caused by the action of FRP is:

$$f_{lf} = E_{fl}\varepsilon_f \tag{4}$$

where ε_f is the tensile strain of FRP and E_{fl} is a measure of the stiffness of the FRP composite or the FRP lateral modulus, which is given by:

$$E_{fl} = \frac{2tE_f}{D} \tag{5}$$

where t is the thickness of the FRP, E_f is the elastic modulus of the FRP, and D is the cylinder diameter.

In addition, no space was required between SR and GFRP tube or concrete cover, which improves the bearing capacity of GFRP–SR confined concrete, as reported by Eid, et $al.$ [33].

Fabrication of Specimens

Unidirectional glass fiber sheets were firstly cut into appropriate lengths for each layer category of GFRP tubes. The GFRP sheets were saturated with epoxy on the surfaces using paintbrushes or rollers and then wrapped around the PVC tubes. The installed tubes had a fiber orientation in the circumferential direction of the cylinders. Two narrow glass fiber sheets were wrapped around both ends of the tubes to avoid premature failure of the specimens. The GFRP tubes were formed and pulled out from PVC tubes after three to four hours, and then dried for seven days. The installed SR together with the GFRP tube was termed as a GFRP–SR tube. Concrete was cast and poured into the GFRP–SR tube.

Material Properties

Concrete

The concrete was designed with a 28 day compressive strength of 30 MPa. The concrete mix design is shown in Table 2. The tested average compressive strength and corresponding strain of the concrete were 30.04 MPa and 0.002, respectively.

Table 2. Concrete mixture proportions.

f_c' (MPa)	W/C	Water (kg/m^3)	Cement (kg/m^3)	Fine Aggregates (kg/m^3)	Coarse Aggregates (kg/m^3)
30	0.51	195.0	382.3	583.3	1239.4

Steel Reinforcement

The mechanical properties of the steel bars were determined using five specimens of the steel bars through the standard tests. The yield strength of the bar was 356 MPa.

FRP Composites

Unidirectional glass FRP sheets with ply thickness of 0.436 mm were used to fabricate the tubes. The mechanical properties of GFRP, including the modulus, tensile strength, and tensile strain, were determined through flat coupon tensile test, in accordance with ASTM D3039-M08 [60], as displayed in Figure 3. Prior to testing, aluminum flat bars were glued to the ends of the coupons to avoid premature failure of the coupon ends. The fiber volume fraction (ρ_f) of the FRP coupons was 25.23% and ρ_f of FRP tubes are provided in Table 1. The measured average tensile modulus, ultimate tensile strength, and tensile strain were 60.8 GPa, 967 MPa, and 0.016, respectively and these properties provided by the manufacturers were 63.0 GPa, 1189 MPa, and 0.019, respectively.

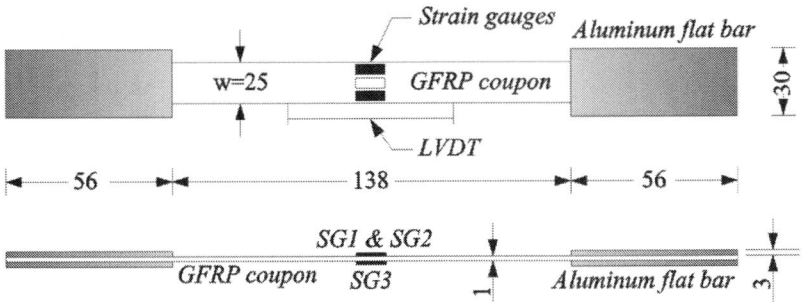

Figure 3. GFRP tension coupon details (unit: mm).

Ductility Index and Energy Consideration for Ductility Index at Failure

A ductility index, μ (Jo *et al.* [61]), to describe ductility is given by:

$$\mu = \frac{1}{2}\left(\frac{E_{tot}}{E_{el}} + 1\right) \tag{6}$$

and

$$E_{tot} = E_{iel} + E_{el} \tag{7}$$

where E_{tot} is the total energy absorbed during deformation; E_{iel} is the inelastic energy absorbed during deformation; and E_{el} is the elastic energy absorbed during deformation (see Figure 4).

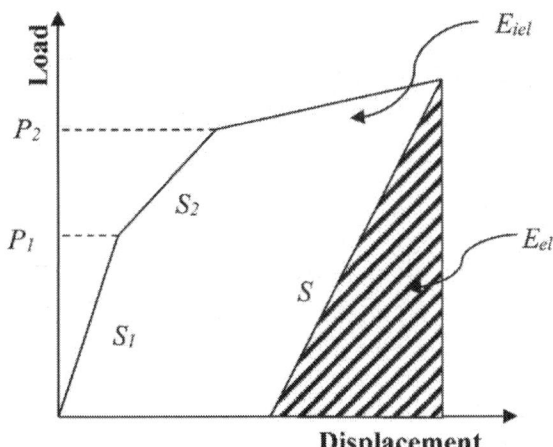

Figure 4. Evaluation method of elasticity energy.

In general, μ given as the ultimate strain divided by yielding strain has been normally used to evaluate the ductility of confined concrete. However, the yield point was difficult to define in this paper because of the slight difference observed in the stress–strain behavior of the steel and FRP composites, such that the conventional definition of the ductility index could not be used.

The E_{el} can be estimated from unloading tests. E_{el} can be computed as the area of the triangle formed at the failure load by the line having the weighted average slope of the two initial straight lines of the load–deflection curve, as illustrated in Figure 4. The slope is given by:

$$S = \frac{P_1 S_1 + (P_2 - P_1)S_2}{P_2} \tag{8}$$

where S_1 and S_2 are the slopes of the two initial straight lines of the load–deflection curve.

In this research, the ductility index was calculated by obtaining an average slope with the same method through curve fitting, as shown in Figure 4.

Test Instrumentation

All of the specimens were tested at the Structural Laboratory of Hunan University using a compression machine (WeiKE Machine, hydraulic, Zaozhuang, China) under stress control mode with a constant rate of 0.20 MPa/s based on ASTM C39 [62]. The acquired data included the applied axial load (P), axial deformation of concrete, transverse and axial strains of the GFRP tube, and the tensile strain of SR. As shown in Figure 5, the axial displacement was measured using four linear variable displacement transducers placed at the middle portion of the cylinders. For each GFRP–SR confined concrete specimen, four hoop strain gauges with a gauge length of 10 mm and four axial strain gauges with a gauge length of 20 mm were installed at the middle portion of the specimen. Two gauges with a gauge length of 5 mm were mounted at the middle portion of SR to measure the tensile strain. Although the GFRP tube was not directly bearing on the loading plates at the ends, some axial stress existed in the GFRP tube because of the bond transfer between tube and concrete.

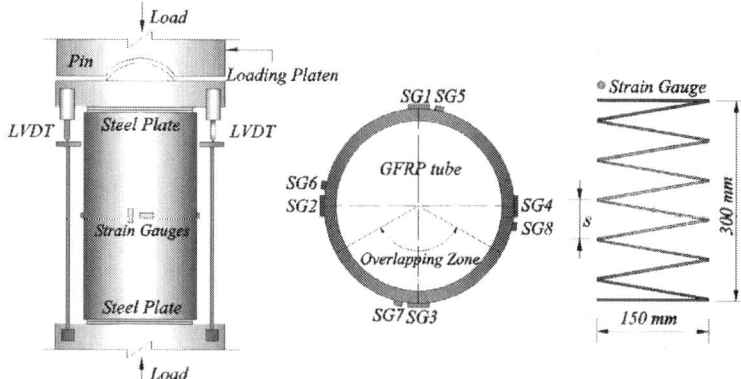

Figure 5. Test set-up and instrumentation configurations.

RESULTS AND DISCUSSION

Failure Modes

The failure modes of GFRP–SR confined concrete cylinder specimens are shown in Figure 6. All of the confined concrete specimens failed by tensile rupture of the GFRP tube in the hoop direction. The failure process was also quiet because of a relatively gradual rupture of the GFRP tube, unlike the explosive process observed in CFRP confined concrete cylinders. The failure situations of all specimens were quite similar. However, for specimens with lower SR volumetric ratio, more core concrete crushed and squeezed out after the rupture of GFRP compared with the higher ones. Moreover, the axial compressive strength and strain of specimens with higher SR volumetric ratio and more layers of GFRP tube were larger, resulting in a higher degree of fragmentation of the concrete core. In general, the GFRP tube and SR showed synchronized transverse deformation at the initial stage of loading and the concrete core was well-preserved when the GFRP tube ruptured because of the confinement of SR. However, the residual compressive behaviors of concrete core confined by SR with different SR volumetric ratios were quite different, *i.e.*, with the increase of compressive load, relatively more core concrete between the larger pitch of SR began to drop out and more apparent axial deformation occurred for core concrete confined with a lower volumetric ratio of SR. Finally, the SR failed with a noise by tearing apart at random points for the core concrete confined with high SR volumetric ratio, but which did not occur to the specimens with relatively low SR volumetric ratio.

Figure 6. Typical failure of specimens.

Axial Stress–Strain Relationships

Axial stress *versus* axial and transverse strain curves of unconfined and confined concrete are shown in Figure 7. In general, the curves can be divided into three stages: the first linear stage, transition zone, and the second linear stage. At the initial stage, the stress–strain responses of all the confined concrete were similar to that of the unconfined concrete, indicating that the confinement of the GFRP–SR tube was not activated. When the axial stress became higher than the unconfined concrete strength, f_c', the concrete lateral strain increased obviously, resulting in the increase of the confinement lateral pressure. Once the GFRP–SR tube was activated to confine the concrete, the curve entered the nonlinear transition region, where considerable micro-cracks appeared in the concrete and led to the lateral expansion of the concrete core. Similar to the conventional FRP-confined concrete with a sufficient level of FRP confinement [18], the stress–strain curves of GFRP–SR confined concrete also exhibited an ascending second linear branch. However, a distinct difference was found between the stress–strain curves of conventional FRP-confined concrete and GFRP–SR confined concrete, *i.e.*, the transition zone of the stress–strain curve of GFRP–SR confined concrete was much longer than that of conventional FRP-confined concrete after reaching the ultimate compressive strength of the unconfined concrete. It is believed this difference is attributed to the stronger confinement provided by the GFRP–SR tube. Unlike conventional FRP-confined concrete, less and slower cracks appeared in the core of GFRP–SR confined concrete after reaching the ultimate compressive strength of the unconfined concrete, which resulted in a less and relatively slow decrease of stiffness.

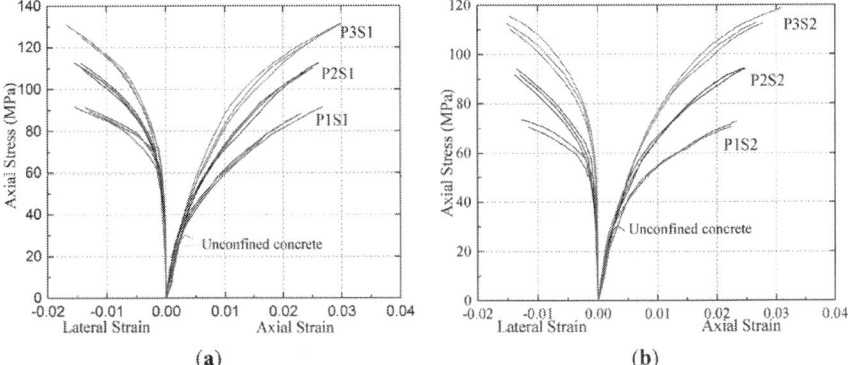

Figure 7. Axial stress *versus* axial and lateral strain curves for GFRP–SR confined cylinder specimens with SR volumetric ratio of (**a**) 3% and (**b**) 1.5%.

Residual Compressive Behavior of Confined Concrete after Rupture of the GFRP Tube

Upon rupture of the GFRP tube, a compressive test was further performed to evaluate the residual compressive behavior of the SR-confined concrete core. The axial compressive stress versus the axial strain curves of SR-confined concrete upon rupture of GFRP are shown in Figure 8. The residual compressive strength of the concrete core with SR confinement was still larger than the ultimate compressive strength of the unconfined concrete. The concrete core with higher SR volumetric ratio showed a larger residual compressive strength; the compressive strengths of specimens P3S2 and P3S1 were 36.8 MPa and 62.2 MPa, respectively. Therefore, the volumetric ratio of SR had a significant influence on the compressive strength of GFRP–SR confined concrete cylinders upon rupture of the GFRP tube. The high residual compressive strength is highly significant in the context of sustaining the residual structure in strong earthquakes. The recent studies by [63,64,65] have focused on the use of polypropylene fiber ropes (PPFRs) as external reinforcement to confine concrete cylinders. Similar to GFRP–SR confined concrete, when used in hybrid confining schemes [64] with GFRP jackets, external PPFRs presented no fracture even after the fracture of the GFRP and furthermore they could be reused. Adequate PPFR confinement could resist the fracture of the FRP jacket and presented an acceptable temporary load drop. Then, further upgrade of the bearing load capacity of the columns followed. Moreover, extremely high ultimate strain can also be achieved for concrete confined by FRP jackets with a large rupture strain, which leads to more ductile behavior and greater energy absorption [6].

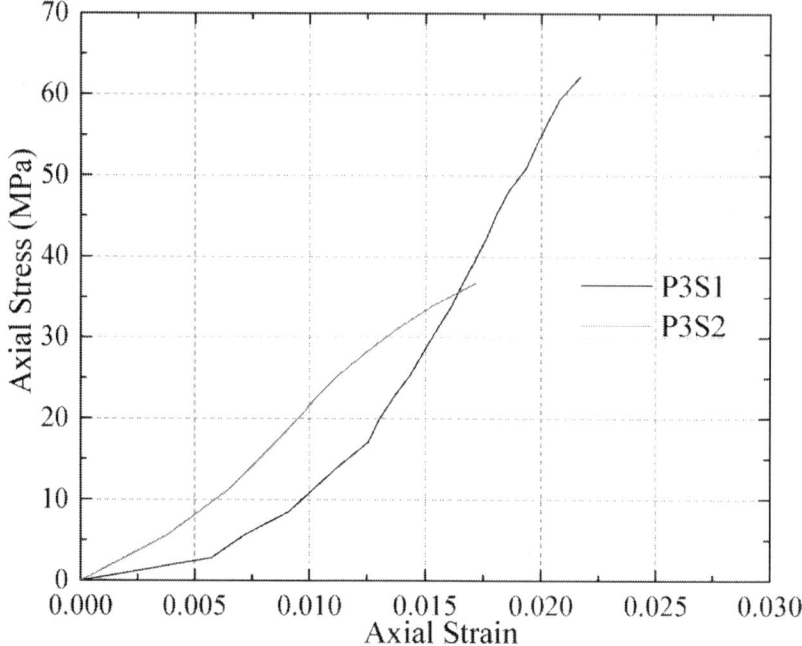

Figure 8. Axial stress *versus* axial strain curves of the reloaded confined concrete.

Ultimate Condition

Table 3 presents a summary of the experimental test results. A significant enhancement of the strength, as well as the ductility for the GFRP–SR confined cylinders, was achieved by increasing the thickness of GFRP tubes and the volumetric ratio of SR. For example, the average compressive strength obtained for specimens P1S2 and P2S1 were 70.95 MPa and 109.71 MPa, respectively, whereas the average ultimate compressive strain determined for specimens P1S2 and P2S1 were 0.0222 and 0.0244, respectively. The ultimate condition shown in Table 3 was dominated by the rupture of the GFRP tube and the rupture of GFRP in specimens with higher volumetric SR ratio, which corresponded to larger axial compressive strength and strain, as reported also by Lee *et al.* [34]. The concrete ultimate axial strains (ε_{cu}) corresponding to failure varied widely from 0.0222 to 0.03, with a tendency to increase for specimens with more GFRP layers and higher SR volumetric ratio. Moreover, the maximum actual lateral confining pressures of SR and GFRP tubes (denoted by $f_{ls,a}$ and $f_{lf,a}$, respectively) are also provided in Table 3. Increasing the volumetric SR ratio for a cylinder specimen with the same FRP confinement results in increased maximum actual lateral confining pressures of GFRP tubes, which was different from the test results of Eid and Paultre (2009) [33] for CFRP-TSR confined concrete. Thus, the presence of SR helped not only in confining the lateral deformation of the core concrete, but also in increasing the confinement action of the GFRP tubes.

The recorded GFRP strains corresponding to failure ($\varepsilon_{fu,a}$) ranged from 0.0121 to 0.0154, which were approximately 75.6%–96.2% of the rupture strains obtained for the tensile coupons (ε_{fu}).

Based on the test results, the difference between the actual FRP rupture strain in FRP–SR-confined concrete specimens and the FRP ultimate tensile strain obtained from a standard tension coupon test can be attributed to the following: (a) the non-uniform deformation of cracked concrete; and (b) the stress state in the GFRP tube was not a strictly pure tension condition as that for the flat coupon tension tests, explained by Matthys et al., (1999) [4], Xiao et al., (2000) [13], De Lorenzis and Tepfers (2003) [43], and Lim and Ozbakkaloglu (2014) [66]. Moreover, according to test results in Table 3, the inelastic energy absorbed by confined concrete cylinders corresponding to failure was much more than the elastic energy absorbed, and the elastic-to-inelastic energy dissipation ratios were relatively constant.

Table 3. Results of confined concrete cylinder specimens (150 mm × 300 mm).

Specimen	f'_c (MPa)	μ	E_c (kJ)	ε'_c	ε_{fu}	E_{tot} (kJ)	ε_{ca}	$\varepsilon_{cu}/\varepsilon'_c$	f_{cu}	f_{cu}/f'_c	$\varepsilon_{fu,a}$	$\frac{\varepsilon_{fu,a}}{\varepsilon_{fu}}$	$f_{lf,a}$ (MPa)	$f_{ls,a}$ (MPa)
P1S1	30.04	5.77	0.51	0.005	0.016	7.58	0.0223	1.46	86.24	2.87	0.0140	0.875	4.95	5.16
P2S1	30.04	5.81	0.51	0.005	0.016	11.03	0.0244	4.88	109.71	3.66	0.0146	0.913	10.32	5.16
P3S1	30.04	6.06	0.51	0.005	0.016	17.91	0.0300	6.00	131.17	4.37	0.0154	0.963	16.32	5.16
P1S2	30.04	5.71	0.51	0.005	0.016	6.56	0.0222	4.44	70.95	2.37	0.0121	0.756	4.27	2.31
P2S2	30.04	5.76	0.51	0.005	0.016	8.23	0.0241	4.82	94.24	3.14	0.0134	0.838	9.48	2.31
P3S2	30.04	5.95	0.51	0.005	0.016	13.30	0.0286	5.72	114.67	3.82	0.0148	0.925	15.70	2.31

Influence of Experiment Variables

SR

Figure 9 shows an indication of the effectiveness of the dual confinement mechanism (SR and GFRP). It shows the relationship and development of the lateral strains in SR and FRP for specimens P1S2 (Figure 9a) and P3S2 (Figure 9b). The initial portions of the curves show that the tensile strain developed in SR, ε_h, was quite similar to the strain developed in the FRP, ε_f. However, at the same level of axial strain after the unconfined concrete strain, ε_c', TSR strain was less affected than FRP strain. Therefore, the inelastic expansion behavior of the concrete had a greater influence on ε_f than on ε_h. This result was also reported by Eid and Paultre [33]. The SR reached its yield strength prior to the rupture of FRP, indicating that at the FRP rupture state, the specimens were subjected to the maximum confining pressures imposed by SR and FRP composite, respectively.

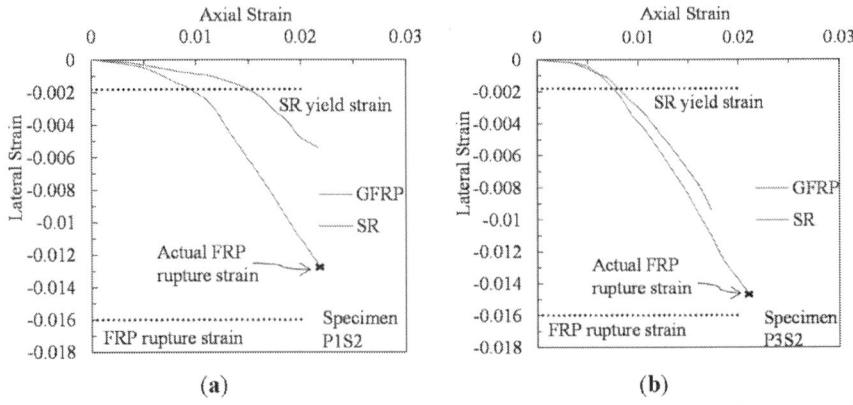

Figure 9. Lateral strain developed in the SR and FRP for specimens (**a**) P1S2 and (**b**) P3S2.

A study by Ozbakkaloglu et al. [67] reviewed and assessed 88 existing FRP-confined concrete models for FRP-confined concrete in circular sections. According to the assessment results, the strength model proposed by Teng et al. [47] is the only analysis-oriented one among the several top performing models. Figure 10 shows the comparison between the analytical axial stress–strain curves of GFRP-confined concrete obtained from Teng et al. [47] and the experimental results of GFRP–SR confined concrete in this paper. As shown in Figure 10, the transition zone of GFRP–SR confined concrete was much longer than that of conventional FRP-confined concrete after reaching the ultimate compressive strength of the unconfined concrete. Moreover, a significant enhancement of strength as well as ductility for the GFRP-confined cylinders was achieved by adding SR, e.g., the compressive strength and strain of specimen P1S2 were 70.95 MPa and 0.0222, respectively, whereas for concrete confined by one layer GFRP, the compressive strength and strain were 48.43 MPa and 0.017, respectively.

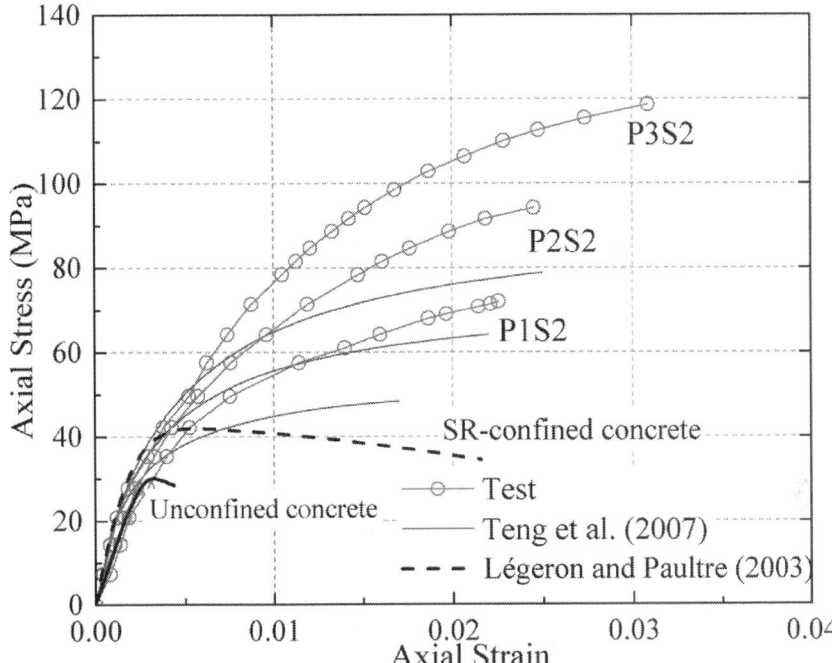

Figure 10. Comparison between axial stress–strain curves of GFRP-confined concrete or SR-confined concrete and the experimental results of GFRP–SR confined concrete.

As shown in Table 3, E_{tot} = 8.23 and 11.03 kJ and E_{tot} = 13.30 and 17.91 kJ were obtained for specimens P2S2 and P2S1 and for specimens P3S2 and P3S1, which had an energy gain of 2.80 and 4.61 kJ, respectively. Thus, the volumetric SR ratio had more significant influence on the energy dissipation of confined concrete with more GFRP layers.

Numbers of FRP Layers
Based on the current test results, the average ratio $\varepsilon_{fu,a}/\varepsilon_{fu}$ increased for the concrete cylinder specimens with higher volumetric SR ratio and more FRP layers (see Table 3), e.g., the average ratios $\varepsilon_{fu,a}/\varepsilon_{fu}$ = 0.875, 0.913, and 0.756 were obtained for specimens P1S1, P2S1 and P1S2, respectively. Moreover, the GFRP layers had a more significant influence on the energy dissipation of confined concrete with higher volumetric SR ratio (e.g., E_{tot} = 11.03 kJ and 17.91 kJ for specimens P2S1 and P3S1 and E_{tot} = 8.23 kJ and 13.30 kJ for specimens P2S2 and P3S2, which were equivalent to energy gain of 6.88 kJ and 5.07 kJ, respectively; Table 3).

The Légeron and Paultre (2003) [48] model, is suitable to represent the axial behavior of circular concrete columns of normal- and high-strength concrete (20–140 MPa) confined by normal- or high-strength (300–1400 MPa) confinement steel. This model, which predicts experimental results with good accuracy [59], is defined by ascending and descending branches, which is the typical behavior of TSR-confined concrete. The comparison between the analytical axial stress–strain curves of SR-confined concrete obtained from Légeron and Paultre (2003) [48] and the experimental results of GFRP–SR confined concrete is showed in Figure 10. As shown in Figure 10, different from SR-confined concrete, the axial stress–strain curve of GFRP–SR confined concrete showed a much longer nonlinear transition zone and an ascending second linear branch, which result in much higher axial ultimate stress.

Axial-Transverse Strain Responses

Recent studies by Lim and Ozbakkaloglu (2015) [68] and Ozbakkaloglu *et al.* (2013) [67] reviewed and assessed a number of existing models on lateral strain-to-axial strain relationship of confined concrete. According to the assessment results, the model proposed by Teng *et al.* (2007) [47] showed a better model performance compared with its counterparts, by adding the influence of TSR, this model was further revised and proposed by Teng *et al.* (2014) [57] for FRP–TSR confined concrete. Moreover, based on a large number of experimental test results of both FRP-confined and actively confined concretes, a generic model was proposed by Lim and Ozbakkaloglu (2015) [68] to describe the lateral strain-to-axial strain relationship of confined concrete, which can predict the trend and critical coordinates of the lateral strain-to-axial strain curves of both FRP-confined and actively confined concretes accurately [68]. The predictions based on models proposed by Lim and Ozbakkaloglu (2015) [68] and Teng *et al.* (2014) [57] are compared with experimental results for the lateral strain-to-axial strain curves of GFRP-SR confined concrete specimens, as illustrated in Figure 11. It is evident from Figure 11 that the direct use of the Lim and Ozbakkaloglu [68] model results in a higher value of slope, v_c', but provides more accurate predictions for initial and transition parts of the curves than Teng's model [57].

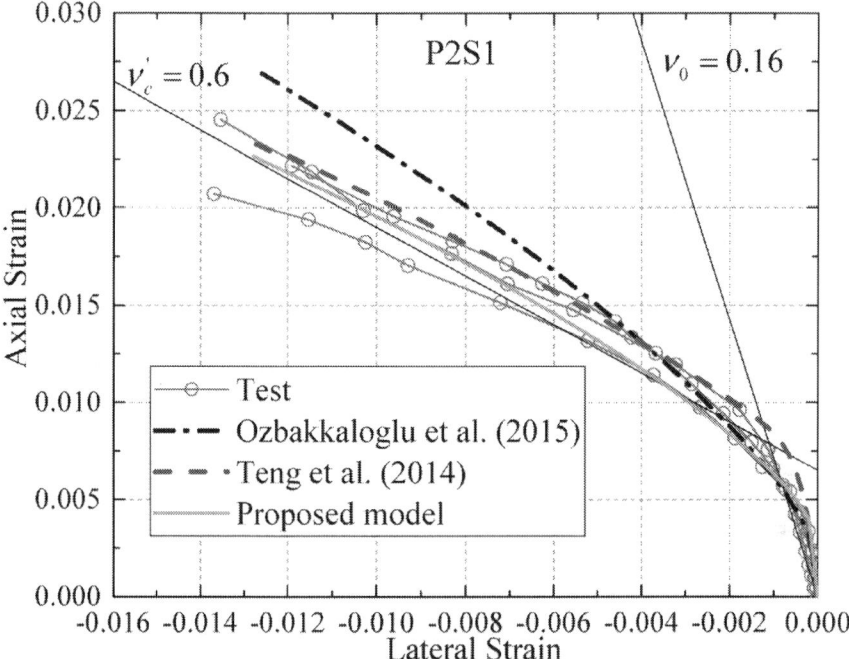

Figure 11. Comparison of model predictions with experimental lateral strain-to-axial strain curves.

The following equations were adopted by Lim and Ozbakkaloglu (2015) [68] for the axial strain–hoop strain relationship:

$$\varepsilon_c = \frac{\varepsilon_h}{\nu_0 \left[1 + \left(\frac{\varepsilon_h}{\nu_0 \varepsilon_c'} \right)^n \right]^{1/n}} + 0.04\varepsilon_h^{0.7} \left[1 + 21 \left(\frac{f_{le}}{f_c'} \right)^{0.8} \right]$$

$$(9)$$

$$\nu_0 = 8 \times 10^{-6} f_c'^2 + 0.0002 f_c' + 0.138$$

$$(10)$$

$$\varepsilon_c' = (-0.067 f_c'^2 + 29.9 f_c' + 1053) \times 10^{-6} \qquad (11)$$

$$n = 1 + 0.03 f_c'$$

(12)

where ε_c is the axial strain, ε_h is the lateral strain, f_{le} is the corresponding confinement pressure for a given lateral strain, v_0 is the initial Poisson's ratio of concrete, f_c' is the peak unconfined concrete strength in MPa, ε_c' is the peak unconfined concrete strain, n is the curve-shape parameter to adjust the initial transition radius of the predicted lateral strain-to-axial strain relationship curve.

In this study, by adding the influence of SR, a revised model based on the expressions proposed by Lim and Ozbakkaloglu (2015) [68] was suggested to describe the lateral strain-to-axial strain relationship curve of GFRP–SR confined concrete, which is given by the following equation:

$$\varepsilon_c = \frac{\varepsilon_h}{v_0 \left[1 + \left(\frac{\varepsilon_h}{v_0 \varepsilon_c'}\right)^n\right]^{1/n}} + 0.04 \varepsilon_h^{0.7} \left[1 + 21\left(\frac{f_{lf}}{f_c'}\right)^{0.8} + 2.6\left(\frac{\alpha f_{ls}}{f_c'}\right)^{0.8}\right]$$

(13)

where $\alpha = 1.59 + 15.1\rho_{ls}$ [57], ρ_{ls} is the ratio between the confining stiffness of the FRP jacket and the effective confining stiffness of SR, which is proposed by Teng *et al.* (2014) [57] to be the parameter to account for the interaction between FRP and SR:

$$\rho_{ls} = \frac{E_f t s d_s}{K_e E_s A_s D}$$

(14)

where E_s, A_s, d_s = elastic modulus, cross-sectional area and diameter of center line of SR.

Figure 11 shows good agreement between the analytical curves obtained from the proposed model and experimental results for three specimens P2S1.

Note that the value of v_c' decreases either with an increase in the yield strength of the steel or with an increase in the elastic modulus of the fiber material. This is consistent with the experimental observations

[13,33,66,69] or the analytical researches [13,59,69,70] for concrete with single or combined confinement.

ANALYTICAL MODELING FOR CONCRETE CONFINED WITH BOTH GFRP TUBES AND SR UNDER MONOTONIC COMPRESSION

The test results indicated that concrete cylinders showed higher strength and ductility when confined with both SR and GFRP composites. Most of the models for confined concrete were based on several researches on concrete confined with one material, and therefore cannot represent the behavior of concrete confined with both SR and FRP. The predictions of the models proposed by Lee et al., (2010) [35], Teng et al., (2014) [57], Chastre and Silva (2010) [58] and Eid and Paultre (2008) [59] are compared with experimental results in Figure 12 for the stress–strain curves of GFRP–SR confined concrete specimens with a low (Specimen P3S2) or medium (Specimen P3S1) confining pressure from SR. The predicted curves all terminate at a point where the experimental FRP rupture strain ($\varepsilon_{fu,a}$) is reached. It is evident from Figure 12 that models proposed by Teng et al., (2014) [57] and Chastre and Silva (2010) [58] are superior to the two other existing models (Lee et al., (2010) [35]; Eid and Paultre (2008) [59]). However, the prediction of the model proposed by Teng et al., (2014) [57] significantly underestimates the responses of GFRP–SR confined concrete specimens at the second linear branches of the stress–strain curves. Moreover, the direct use of the Chastre and Silva (2010) [58] model result in higher ultimate axial stresses for the GFRP–SR confined concrete specimens, but provides reasonable predictions for the types of stress-strain curves. Since the model described in Chastre and Silva (2010) [58] was developed based on several researches on CFRP retrofitted circular columns with 150–400 mm diameter and H/D between 3 and 5, it can be further modified to predict the stress-strain curve of GFRP–SR confined concrete based on tests results.

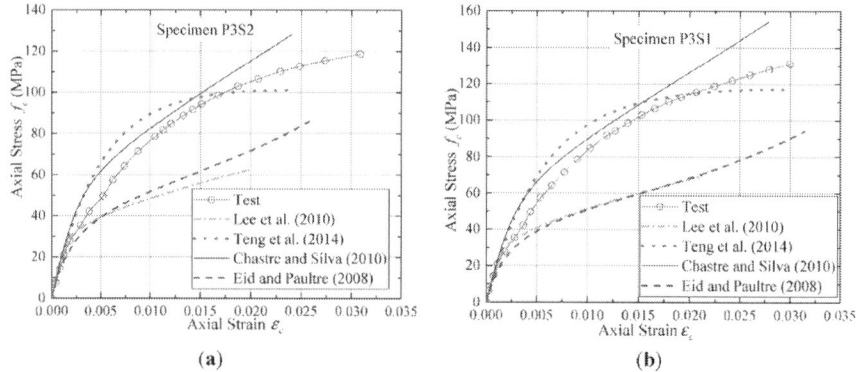

Figure 12. Performance of models for GFRP–SR confined concrete with a (**a**) low (**b**) medium level of steel confinement.

Proposed Stress Equations

For the monotonic actions, The compressive strength of confined concrete can be expressed in the following common form [51,52,67,71]:

$$f_{cu} = f_D + k_1 f_{le} \tag{15}$$

Equation (15) includes the contribution of the confinement given by the lateral steel reinforcement and by the FRP tube, and this equation considers superposed effects at rupture. Moreover, Equation (15) was calibrated for GFRP and SR ($k_1 = 4.6$) through experimental tests (Figure 13).

The compressive strength of the concrete cylinder, f_D [72], can be given by [48,67,70]:

$$f_D = \alpha f_c' \tag{16}$$

where α is the scale effects coefficient (($1.5 + D/H$)/2 = 1 in this paper), and D and H are the diameter and the height of the cylinder, respectively.

Assuming the aforementioned superposed effects at rupture of the FRP tube (f_{lf}) and the steel spirals (f_{ls}) confinement, the lateral confining pressure (f_{le}) is defined by:

$$f_{le} = f_{lf} + f_{ls} \tag{17}$$

where f_{ls} and f_{lf} are given by Equations (3) and (4), respectively, and $f_h = f_{hy}$ [48].

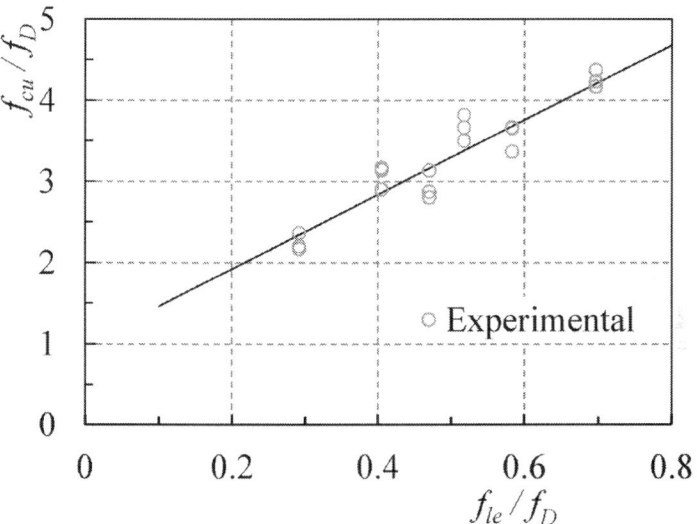

Figure 13. Relationship between f_{cu}, f_D and f_{le} of concrete cylinders confined with GFRP and SR.

The axial strain in rupture (ε_{cu}) can be expressed in the following form [54]:

$$\varepsilon_{cu} = k_2 \varepsilon_{c0} \left(\frac{f_{le}}{f_D} \right)^{0.7} \tag{18}$$

where ε_{c0} is adapted from Eurocode 2 (2004) [73]:

$$\varepsilon_{c0} = \frac{0.7}{1000} (f_c')^{0.31} \tag{19}$$

Equation (18) was obtained for GFRP and SR ($k_2 = 20$) by regression of experimental data (Figure 14) of concrete cylinders confined with GFRP tubes and SR.

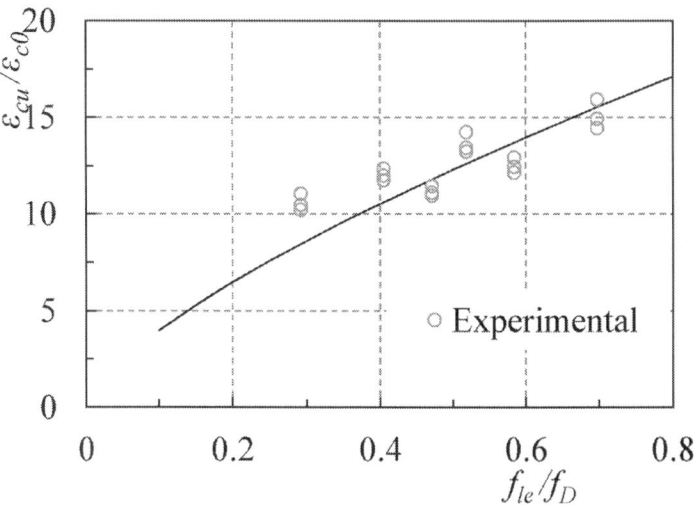

Figure 14. Relationship between $\varepsilon_{cu}/\varepsilon_{c0}$ and f_{le}/f_D of concrete cylinders confined with GFRP and SR.

Proposed Stress–Strain Model for FRP–SR Confined Concrete in Compression

Note that the model proposed for the GFRP–SR confined concrete stress–strain curve of cylinders subjected to monotonic axial compression was based on the stress–strain law depicted in Figure 15.

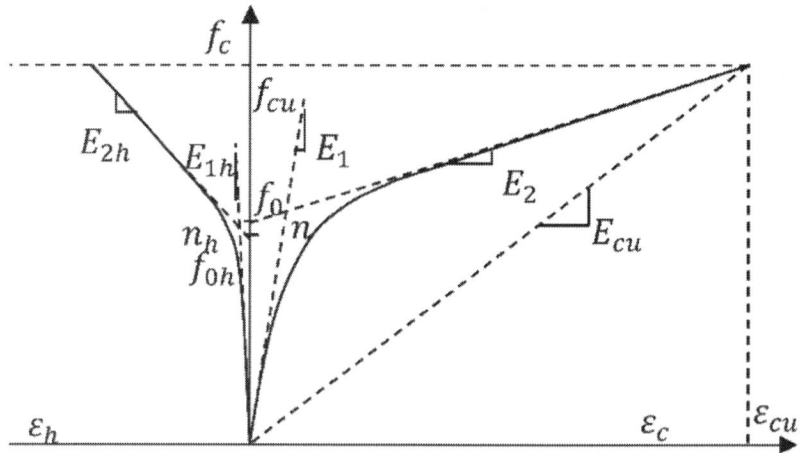

Figure 15. Proposed stress–strain model for GFRP–SR confined concrete in compression.

For the monotonic actions, the stress–axial strain relationship was bi-linear for the concrete confined with GFRP and SR (Figure 15) and was based on a versatile expression of four parameters (E_1, E_2, f_0, n), as initially proposed by Richard and Abbott [74]:

$$f_c = \frac{(E_1 - E_2)\varepsilon_c}{\left[1 + \left(\frac{(E_1 - E_2)\varepsilon_c}{f_0}\right)^n\right]^{\frac{1}{n}}} + E_2\varepsilon_c \leq f_{cu} \tag{20}$$

With the parameters calibrated according to available experimental results:

$$\begin{cases} E_1 = 3950\sqrt{f_D} & (a) \\ E_2 = 0.73E_{cu}\sqrt{\frac{f_{le}}{f_D}} & (b) \\ f_0 = f_D + 1.75f_{le} & (c) \end{cases} \tag{21}$$

E_{cu} can be estimated applying Equations (15) and (18) to the following expression:

$$E_{cu} = \frac{f_{cu}}{\varepsilon_{cu}} \tag{22}$$

The axial stress–lateral strain curve was also bi-linear:

$$f_c = \frac{(E_{1h} - E_{2h})\varepsilon_h}{\left[1 + \left(\frac{(E_{1h} - E_{2h})\varepsilon_h}{f_{0h}}\right)^{n_h}\right]^{\frac{1}{n_h}}} + E_{2h}\varepsilon_h \leq f_{cu} \tag{23}$$

with

$$\begin{cases} E_{1h} = \frac{E_1}{\nu} & (a) \\ E_{2h} = 150(f_{le})^{1.16}(f_D)^{-0.16} & (b) \\ f_{0h} = 1.25f_D + 2.3f_{le} & (c) \end{cases} \tag{24}$$

For the axial stress–axial strain curve, the slope of the first branch was considered essentially, followed by the curves of the unconfined concrete, given that the GFRP–SR tube had a passive behavior and was only activated for a stress level similar to the maximum stress of the unconfined concrete, which was defined by Equation (21a) [1,54,58].

The slope of the second branch, E_2 (Equation (21b)), was experimentally calibrated (Figure 16) with the function of the slope of confinement of several concrete cylinders confined with GFRP tubes and SR. The stress value f_0 (Equation (21c)) and the parameter $n=2$ can be estimated based on the calibration of curves through the experimental tests (Figure 17).

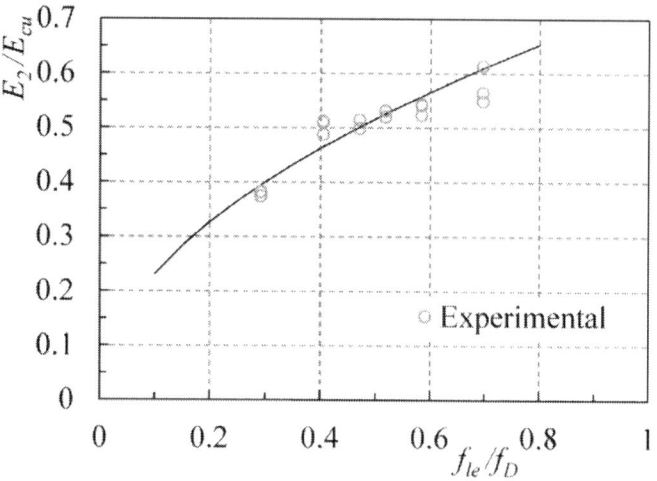

Figure 16. Parameter E_2-experimental calibration with concrete cylinders confined with GFRP tubes and SR.

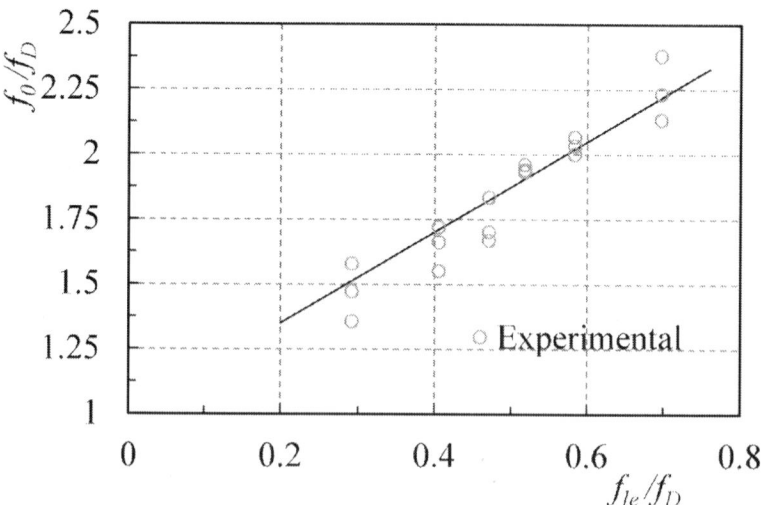

Figure 17. Parameter f_0-experimental calibration with concrete cylinders confined with GFRP tubes and SR.

For the axial stress–lateral strain curve, the slope of the first branch was dependent on the concrete Poisson's ratio ($v = 0.16$), which is given by Equation (24a). The parameter n_h was assumed to be 2. The slope of second branch (Equation (24b)) and the parameter f_{0h}(Equation (24c)) were determined after experimental calibration, as shown in Figure 18 and Figure 19.

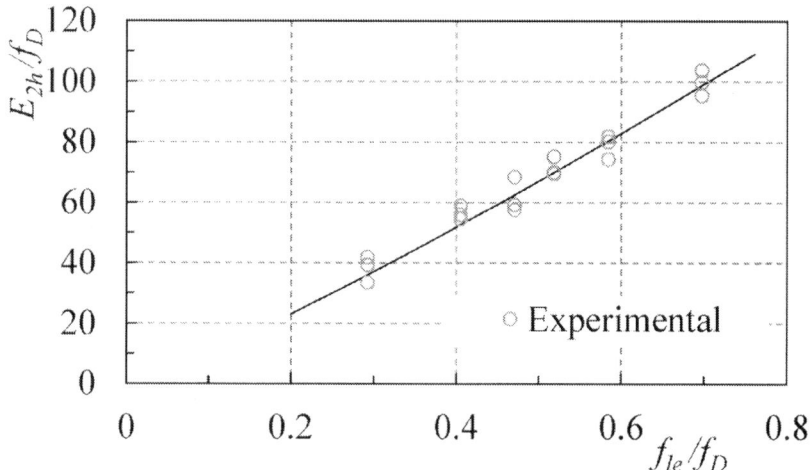

Figure 18. Parameter E_{2h}-experimental calibration with concrete cylinders confined with GFRP tubes and SR.

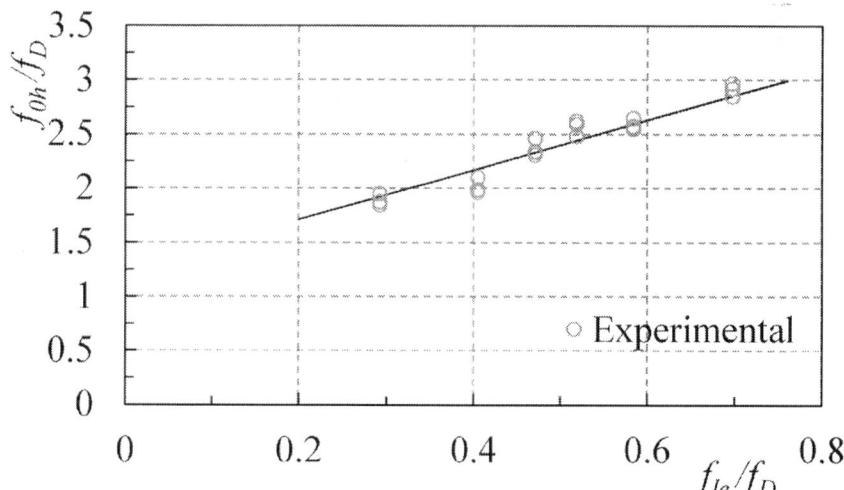

Figure 19. Parameter f_{0h}-experimental calibration with concrete cylinders confined with GFRP tubes and SR.

Comparison of the Model Proposed with the Experimental Results

Figure 20 shows good agreement between the analytical curves obtained from the proposed stress–strain model and experimental results for three specimens P1S1, P2S1, and P3S1. Moreover, the predictions of the proposed model are compared in Figure 21 for the stress–strain curves of FRP–TSR confined concrete with experimental results from Chastre and Silva (2010) [58] and Benzaid *et al.*, (2010) [18], which were not used in the development of the proposed model. Only typical comparisons are shown in Figure 21, but those for other specimens are similar. Figure 21 shows that the proposed model is capable of providing accurate predictions for the stress–strain curves.

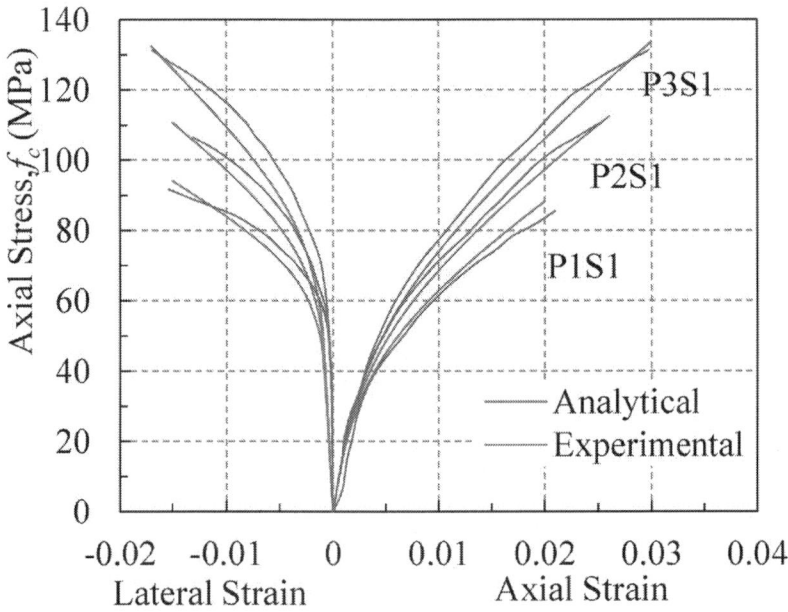

Figure 20. Comparison of compressive stress–strain curves between the predictions and test results for Specimens P1S1, P2S1, and P3S1.

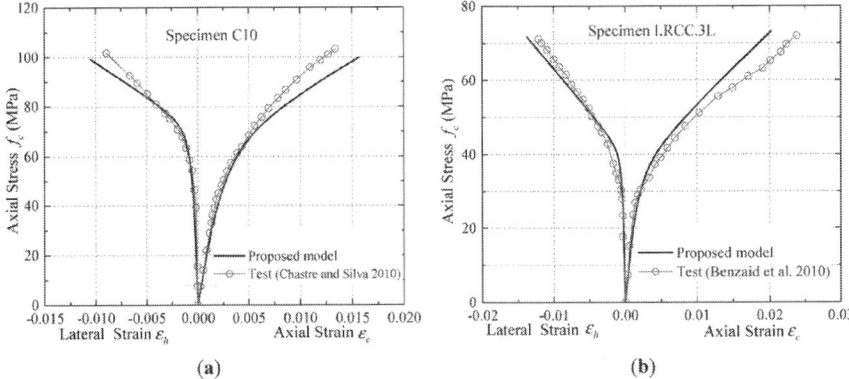

Figure 21. Performance of proposed model for FRP–TSR confined concrete: (**a**) Specimen C10 from Chastre and Silva (2010) [58]; (**b**) Specimen I.RCC.3L from Benzaid *et al.*, (2010) [18].

CONCLUSIONS

1. Significant increase in strength and ductility of concrete can be achieved by using GFRP tubes and SR. Unlike the explosive process observed in CFRP confined concrete cylinders, the failure process of GFRP–SR confined concrete was quiet and the GFRP–SR confined concrete had a good residual compressive strength after the rupture of GFRP.

2. Increasing the volumetric SR ratio for a cylinder specimen with the same FRP confinement results in increased maximum actual lateral confining pressures of GFRP tubes, which is different from the test results of Eid and Paultre (2009) [33] for CFRP–TSR confined concrete.

3. The stress–strain performances of concrete confined with GFRP tube and SR exhibited an ascending bilinear shape with a long transition zone around the stress level of unconfined concrete strength. A model was proposed to describe the relationships between the axial stress–axial strain and axial stress–lateral strain; this model showed good agreement with the experimental results.

4. The test results were compared with predictions of some existing models. For GFRP–SR confined concrete, models proposed by Teng *et al.*, (2014) [57] and Chastre and Silva (2010) [58] are superior to the two other existing models (Lee *et al.*, 2010 [35]; Eid and Paultre 2008 [59]). The direct use of the Chastre and Silva

(2010) [58] model significantly overestimates the ultimate axial stresses of the GFRP–SR confined concrete specimens, but provides reasonable predictions for the types of stress–strain curves.

5. The inelastic energy absorbed by confined concrete cylinders corresponding to failure was much more than the elastic energy absorbed, and the elastic-to-inelastic energy dissipation ratios were relatively constant.

ACKNOWLEDGMENTS

The tests on concrete cylinders confined by GFRP–SR tubes were funded by Hunan Provincial Natural Science Foundation of China (No. 2015JJ1004), National Basic Research Program of China (973 program, Project No. 2012CB026200), and the Sci-Tech Support Plan of Hunan Province (Project No. 2014WK2026). The authors would like to thank Yan Xiao and Giorgio Monti for their valuable input. Special thanks to graduate students Peng Yin and Kai Huang for their help in preparing the specimens and conducting the tests.

AUTHOR CONTRIBUTIONS

Liang Huang and Xiaoxun Sun conceived and designed the experiments; Liang Huang and Xiaoxun Sun performed the experiments; Liang Huang, Xiaoxun Sun and Libo Yan analyzed the data; Liang Huang and Deju Zhu contributed reagents/materials/analysis tools; Liang Huang, Xiaoxun Sun and Libo Yan wrote the paper.

REFERENCES

1. Ahmad, S.M.; Shah, S.P. Stress–strain curves of concrete confined by spiral reinforcement. *ACI Struct. J.* 1982, *79*, 484–490.

2. Mander, J.B.; Priestley, M.J.N.; Park, R. Observed stress strain behavior of confined concrete. *ASCE J. Struct. Eng.* 1988, *114*, 1827–1849.

3. Parvin, A.; Brighton, D. FRP composites strengthening of concrete columns under various loading conditions. *Polymers* 2014, *6*, 1040–1056.
4. Matthys, S.; Taerwe, L.; Audenaert, K. Tests on axially loaded concrete columns confined by fiber reinforced polymer sheet wrapping. *ACI Spec. Publ.* 1999, *188*, 217–228.
5. Lam, L.; Teng, J.G. Ultimate condition of fiber reinforced polymer-confined concrete. *J. Compos. Constr.* 2004, *8*, 539–548.
6. Dai, J.G.; Bai, Y.L.; Teng, J.G. Behavior and modeling of concrete confined with FRP composites of large deformability. *J. Compos. Constr.* 2011, *15*, 963–973.
7. Ilki, A.; Peker, O.; Karamuk, E.; Demir, C.; Kumbasar, N. FRP retrofit of low and medium strength circular and rectangular reinforced concrete columns. *J. Mater. Civil Eng.* 2008, *20*, 169–188.
8. Ozcan, O.; Binici, B.; Ozcebe, G. Seismic strengthening of rectangular reinforced concrete columns using fiber reinforced polymers. *Eng. Struct.* 2010, *32*, 964–973.
9. Wu, Y.F.; Wei, Y.Y. Effect of cross-sectional aspect ratio on the strength of CFRP-confined rectangular concrete columns. *Eng. Struct.* 2010, *32*, 32–45.
10. Ozbakkaloglu, T.; Akin, E. Behavior of FRP-confined normaland high-strength concrete under cyclic axial compression. *J. Compos. Constr.* 2012, *16*, 451–463.
11. Wang, Z.Y.; Wang, D.Y.; Smith, S.T.; Lu, D.G. CFRPconfined square RC columns. I: Experimental investigation. *J. Compos. Constr.* 2012, *16*, 150–160.
12. Xiao, Y.; Wu, H. Compressive behavior of concrete confined by various types of FRP composite jackets. *J. Reinf. Plast. Compos.* 2003, *22*, 1187–1201.
13. Xiao, Y.; Wu, H. Compressive behavior of concrete confined by carbon fiber composite jackets. *J. Mater. Civil Eng.* 2000, *12*, 139–146.
14. Priestley, M.J.N.; Seible, F. *Seismic Assessment and Retrofit of Bridges*; Structural Systems Researck Project; Report No. SSRP 91103. University of California: San Diego, CA, USA; July; 1991; p. 418.
15. Seible, F.; Hegemier, G.A.; Innamorato, D. Developments in bridge column jacketing using advance composites. In Proceedings of the National Seismic Conference on Bridges and Highways, Federal Highway Administration and California Department of Transportation, San Diego, CA, USA, 10–13 December 1995.
16. Ma, R.; Xiao, Y.; Li, K.N. Full-scale testing of a parking structure column retrofitted with carbon fiber reinforced composites. *J. Constr. Build. Mater.* 2000, *14*, 63–71.

17. Teng, J.G.; Chen, J.F.; Smith, S.T.; Lam, L. *RC Structures Strengthened with FRP Composites*; Research Centre for Advanced Technology in Structural Engineering, Department of Civil and Structural Engineering. The Hong Kong Polytechnic University: Hong Kong, China; December; 2000; p. 134.

18. Benzaid, R.; Mesbah, H.; Chikh, N.E. FRP-confined concrete cylinders: Axial compression experiments and strength model. *J. Reinf. Plast. Compos.* 2010, *29*, 2469–2488.

19. Xie, T.; Ozbakkaloglu, T. Behavior of steel fiber-reinforced high-strength concrete-filled FRP tube columns under axial compression. *Eng. Struct.* 2015, *90*, 158–171.

20. Vincent, T.; Ozbakkaloglu, T. Influence of fiber orientation and specimen end condition on axial compressive behavior of FRP-confined concrete. *Constr. Build. Mater.* 2013, *47*, 814–826.

21. Vincent, T.; Ozbakkaloglu, T. Influence of shrinkage on compressive behavior of concrete-filled FRP tubes: An experimental study on interface gap effect. *Constr. Build. Mater.* 2015, *75*, 144–156.

22. Vincent, T.; Ozbakkaloglu, T. Compressive behavior of prestressed high-strength concrete-filled aramid FRP tube columns: Experimental observations. *J. Compos. Constr.* 2015, *2015*.

23. Vincent, T.; Ozbakkaloglu, T. Influence of slenderness on stress–strain behavior of concrete-filled FRP tubes: Experimental study. *J. Compos. Constr.* 2014, *19*, 04014029.

24. Ozbakkaloglu, T.; Vincent, T. Axial compressive behavior of circular high-strength concrete-filled FRP tubes. *J. Compos. Constr.* 2014, *18*, 04013037.

25. Yan, L.; Chouw, N. Dynamic and static properties of flax fibre reinforced polymer tube confined coir fibre reinforced concrete. *J. Compos. Mater.* 2014, *48*, 1595–1610.

26. Ozbakkaloglu, T. Behavior of square and rectangular ultra high-strength concrete-filled FRP tubes under axial compression. *Compos. B Eng.* 2013, *54*, 97–111.

27. Yan, L.; Chouw, N.; Jayaraman, K. Effect of column parameters on flax FRP confined coir fibre reinforced concrete. *Constr. Build. Mater.* 2014, *55*, 299–312.

28. Zhang, B.; Yu, T.; Teng, J. Behavior of concrete-filled FRP tubes under cyclic axial compression. *J. Compos. Constr.* 2014, *2014*.

29. Zhao, J.L.; Yu, T.; Teng, J.G. Stress–strain behavior of FRP-confined recycled aggregate concrete. *J. Compos. Constr.* 2014, *2014*.

30. Monti, G. Confining reinforced concrete with FRP: Behavior and modeling. *Compos. Constr.* 2002, *2002*, 213–222.

31. Richart, F.E.; Brandtzæg, A.; Brown, R.L. *Failure of Plain and Spirally Reinforced Concrete in Compression*; Bulletin No. 190. Engineering Experiment Station, University of Illinois: Urbana, IL, USA, 1929.

32. Kumutha, R.; Palanichamy, M.S. Investigation of reinforced concrete columns confined using glass fiber-reinforced polymers. *J. Reinf. Plast. Compos.* 2006, *25*, 1669–1678.

33. Eid, R.; Roy, N.; Paultre, P. Normal- and high-strength concrete circular elements wrapped with FRP composites. *J. Compos. Constr.* 2009, *13*, 113–124.

34. Triantafyllou, G.G.; Rousakis, T.C.; Karabinis, A.I. Axially loaded reinforced concrete columns with a square section partially confined by light GFRP straps. *J. Compos. Constr.* 2014, *19*, 04014035.

35. Lee, J.Y.; Yi, C.K.; Jeong, H.S.; Kim, S.W.; Kim, J.K. Compressive response of concrete confined with steel spirals and FRP composites. *J. Compos. Mater.* 2010,*44*, 481–504.

36. Demers, M.; Neale, K. Confinement of reinforced concrete columns with fibre-reinforced composite sheets—An experimental study. *Can. J. Civil Eng.* 1999, *26*, 226–241.

37. Matthys, S.; Toutanji, H.; Audenaert, K.; Taerwe, L. Axial load behavior of large-scale columns confined with fiber-reinforced polymer composites. *ACI Struct. J.*2005, *102*, 258–267.

38. Carey, S.A.; Harries, K.A. Axial behavior and modeling of confined small-, medium-, and large-scale circular sections with carbon fiber-reinforced polymer jackets. *ACI Struct. J.* 2005, *102*, 596–604.

39. Rousakis, T.C.; Karabinis, A.I. Adequately FRP confined reinforced concrete columns under axial compressive monotonic or cyclic loading. *Mater. Struct.* 2012,*45*, 957–975.

40. De Luca, A.; Nardone, F.; Matta, F.; Nanni, A.; Lignola, G.P.; Prota, A. Structural evaluation of full-scale FRP-confined reinforced concrete columns. *J. Compos. Constr.* 2010, *15*, 112–123.

41. Tamuzs, V.; Valdmanis, V.; Gylltoft, K.; Tepfers, R. Behavior of CFRP-confined concrete cylinders with a compressive steel reinforcement. *Mech. Compos. Mater.*2007, *43*, 191–202.

42. Wang, Y.C.; Restrepo, J.I. Investigation of concentrically loaded reinforced concrete columns confined with glass fiber-reinforced polymer jackets. *ACI Struct. J.* 2001,*98*, 377–385.

43. De Lorenzis, L.; Tepfers, R. Comparative study of models on confinement of concrete cylinders with fiber-reinforced polymer composites. *J. Compos. Constr.*2003, *7*, 219–237.

44. Teng, J.G.; Lam, L. Behavior and modeling of fiber reinforced polymer-confined concrete. *J. Struct. Eng.* 2004, *130*, 1713–1723.

45. Samaan, M.; Mirmiran, A.; Shahawy, M. Model of concrete confined by fiber composites. *ASCE J. Struct. Eng.* 1998, *124*, 1025–1031.

46. Spoelstra, M.R.; Monti, G. FRP-confined concrete model. *J. Compos. Constr.* 1999, *3*, 143–150.

47. Teng, J.G.; Huang, Y.L.; Lam, L.; Ye, L.P. Theoretical model for fiber-reinforced polymer-confined concrete. *J. Compos. Constr.* 2007, *11*, 201–210.

48. Légeron, F.; Paultre, P. Uniaxial confinement model for normal- and high-strength concrete columns. *J. Struct. Eng.* 2003, *129*, 241–252.

49. Lignola, G.P.; Prota, A.; Manfredi, G. Simplified modeling of rectangular concrete cross-sections confined by external FRP wrapping. *Polymers.* 2014, *6*, 1187–1206.

50. Kawashima, K.; Hosotani, M.; Yoneda, K. Carbon fiber sheet retrofit of reinforced concrete bridge piers. In Proceedings of Workshop on Annual Commemoration of Chi-Chi Earthquake, National Center for Research on Earthquake Engineering, Taipei, Taiwan, 18 September 2000; pp. 124–135.

51. Matthys, S. Structural Behavior and Design of Concrete Members Strengthened with Externally Bonded FRP. Ph.D. Thesis, Ghent University, Gent, Belgium, 2000; p. 345.

52. Lam, L.; Teng, J.G. Design-oriented stress–strain model for FRP-confined concrete. *Constr. Build. Mater.* 2003, *17*, 471–489.

53. Matthys, S.; Toutanji, H.; Taerwe, L. Stress–strain behaviour of large-scale circular columns confined with FRP composites. *J. Struct. Eng.* 2006, *132*, 123–133.

54. Mander, J.B.; Priestley, M.J.N.; Park, R. Theoretical stress–strain model for confined concrete. *J. Struct. Eng.* 1988, *114*, 1804–1826.

55. Saafi, M.; Toutanji, H.; Li, Z. Behavior of concrete columns confined with fiber reinforced polymer tubes. *ACI Mater. J.* 1999, *96*, 500–509.

56. Toutanji, H. Stress–strain characteristics of concrete columns externally confined with advanced fiber composite sheets. *ACI Mater. J.* 1999, *96*, 397–404.

57. Teng, J.G.; Lin, G.; Yu, T. Analysis-oriented stress–strain model for concrete under combined FRP-steel confinement. *J. Compos. Constr.* 2014, *2014*.

58. Chastre, C.; Silva, M.A. Monotonic axial behavior and modelling of RC circular columns confined with CFRP. *Eng. Struct.* 2010, *32*, 2268–2277.

59. Eid, R.; Paultre, P. Analytical model for FRP-confined circular reinforced concrete columns. *J. Compos. Constr.* 2008, *12*, 541–552.

60. *Standard Test Method for Tensile Properties of Polymer Matrix Composite Materials*; ASTM D3039/D 3039M. American Society for Testing and Materials (ASTM): West Conshohocken, PA, USA, 2008.

61. Jo, B.-W.; Tae, G.-H.; Kwon, B.-Y. Ductility evaluation of prestressed concrete beams with CFRP tendons. *J. Reinf. Plast. Compos.* 2004, *23*, 843–859.

62. *Standard Test Methods for Compressive Strength of Cylindrical Concrete Specimens*; ASTM C39. American Society for Testing and Materials (ASTM): West Conshohocken, PA, USA, 2010.

63. Rousakis, T.C. Elastic fiber ropes of ultrahigh-extension capacity in strengthening of concrete through confinement. *J. Mater. Civil Eng.* 2013, *26*, 34–44.

64. Rousakis, T.C. Hybrid confinement of concrete by fiber-reinforced polymer sheets and fiber ropes under cyclic axial compressive loading. *J. Compos. Constr.* 2013, *17*, 732–743.

65. Rousakis, T.C.; Tourtouras, I.S. RC columns of square section—Passive and active confinement with composite ropes. *Compos. B Eng.* 2014, *58*, 573–581.

66. Lim, J.C.; Ozbakkaloglu, T. Hoop strains in FRP-confined concrete columns: Experimental observations. *Mater. Struct.* 2014, *2014*, 1–16.

67. Ozbakkaloglu, T.; Lim, J.C.; Vincent, T. FRP-confined concrete in circular sections: Review and assessment of stress–strain models. *Eng. Struct.* 2013, *49*, 1068–1088.

68. Lim, J.; Ozbakkaloglu, T. Lateral strain-to-axial strain relationship of confined concrete. *J. Struct. Eng.* 2015, *141*, 04014141.

69. Lim, J.; Ozbakkaloglu, T. Factors influencing hoop rupture strains of FRP-confined concrete. *Appl. Mech. Mater.* 2014, *501*, 949–953.

70. Ozbakkaloglu, T.; Lim, J.C. Axial compressive behavior of FRP-confined concrete: Experimental test database and a new design-oriented model. *Compos. B Eng.* 2013, *55*, 607–634.

71. Lam, L.; Teng, J.G. Strength models for fiber-reinforced plastic-confined concrete. *J. Struct. Eng.* 2002, *128*, 612–623.

72. *Concrete Core Testing for Strength*; CS-11. Concrete Society: London, UK, 1976.

73. *Eurocode 2: Design of Concrete Structures: Part 1–1: General Rules and Rules for Buildings*; British Standards Institution: Brussels, Belgium, 2004.

74. Richard, R.M.; Abbott, B.J. Versatile elastic–plastic stress–strain formula. *J. Eng. Mech. Division* 1975, *101*, 511–515.

CITATION

Liang Huang, Xiaoxun Sun, Libo Yan and Deju Zhu, Compressive Behavior of Concrete Confined with GFRP Tubes and Steel Spirals, doi:10.3390/polym7050851

CHAPTER 5

On Displacement-Based and Mixed-Variational Equivalent Single Layer Theories for Modelling Highly Heterogeneous Laminated Beams

R.M.J. Groh and P.M. Weaver

Advanced Composites Centre for Innovation and Science, University of Bristol, Queen's Building, University Walk, Bristol BS8 1TR, UK

ABSTRACT

The flexural response of laminated composite and sandwich beams is analysed using the notion of modelling the transverse shear mechanics with an analogous mechanical system of springs in series combined with a system of springs in parallel. In this manner a zig-zag function is derived that accounts for the geometric and constitutive heterogeneity of the multi-layered beam similar to the zig-zag function in the refined zig-zag theory (RZT) developed by Tessler et al. (2007). Based on this insight a new equivalent single layer formulation is developed using the principle of virtual displacements. The theory overcomes the problem in the RZT framework of modelling laminates with Externally Weak Layers but is restricted to laminates with zero B-matrix terms. Second, the RZT zig-zag function is implemented in a third-order theory based on the Hellinger–Reissner mixed variational framework. The advantage of the Hellinger–Reissner formulation is that both in-plane and transverse stress fields are captured to within 1% of Pagano's 3D elasticity solution without the need for additional stress recovery steps, even for highly heterogeneous laminates. A variant of the Hellinger–Reissner formulation with Murakami's zig-zag function increases the percentage error by an order of magnitude for highly heterogeneous laminates. Corresponding formulations using the Reissner Mixed Variational Theory

(RMVT) show that the independent model assumptions for transverse shear stresses in this theory may be highly inaccurate when the number of layers exceeds three. As a result, the RMVT formulations require extra post-processing steps to accurately capture the transverse stresses. Finally, the relative influence of the zig-zag effect on different laminates is quantified using two non-dimensional parameters.

INTRODUCTION

The application of multi-layered composite materials in load-bearing structures is finding widespread application particularly in the aeronautical, marine and renewable energy industries. Reasons include their high specific strength and stiffness, good fatigue resistance and enhanced design freedom on a micro- and macromechanical level. Furthermore, as material costs reduce the use of carbon fibre composites in large-scale automotive applications is expected to grow considerably in the coming years (Lucintel, 2013).

The design of primary load-bearing structures requires accurate tools for stress analysis. When used around areas of stress concentration or in fail-safe design frameworks composite laminates are often designed to have thicker cross-sections. Under these circumstances non-classical effects, such as transverse shear and normal deformation become important factors in the failure event. These considerations mean that Euler–Bernoulli beam and Kirchhoff plate/shell models that underpin Classical Laminate Analysis (CLA) (Jones, 1998) inaccurately predict global and local deformations. Transverse shear deformations are particularly pronounced in composite materials because the ratio of longitudinal to shear modulus is approximately one order of magnitude larger than for isotropic materials $(E_{iso}/G_{iso} = 2.6, \ E_{11}/G_{xz} \approx 140/5 = 28)$. The analysis of layered composites is also more cumbersome due to transverse anisotropy, and interlaminar continuity (IC) conditions on displacement, transverse shear and transverse normal stress fields.

Most notably, transverse anisotropy, i.e. the difference in layerwise transverse shear and normal moduli, leads to sudden changes in slope of the three displacement fields $u_x, \ u_y, \ u_z$ at layer interfaces. This is known as the zig-zag (ZZ) phenomenon. In fact,Carrera (2001) points out that "compatibility and equilibrium, i.e., ZZ and IC, are strongly connected to each other." Thus, while IC of the displacements requires $u_x, \ u_y, \ u_z$ to beC^0 continuous at interfaces, IC of the transverse stresses forces the displacement fields to be C^1 *discontinuous*. Motivated by these considerations, Demasi (2012) showed that the ZZ form of the in-plane displacements u_x, u_y and u_z can be derived directly from τ_{xz}, τ_{yz} and σzz continuity, respectively. Therefore, an accurate model for multi-layered composite and sandwich structures should ideally address the modelling issues named C_z^0-requirements by Carrera, 2002 and Carrera, 2003b:

1. Through-thickness z-continuous displacements and transverse stresses i.e. the IC condition.
2. Discontinuous first derivatives of displacements between layers with different mechanical properties i.e. the zig-zag effect.

For this purpose high-fidelity 3D Finite Element (FE) models are often employed for accurate structural analysis. However, these models can become computationally prohibitive when employed for laminates with large number of layers, in optimisation studies, for non-linear problems that require iterative solution techniques or for progressive failure analyses. Thus, with the aim of developing rapid, yet robust design tools for industrial purposes there remains a need for further efficient modelling techniques. In this regard particular focus is required on equivalent single layer (ESL) theories because the number of unknowns in these formulations is independent of the number of layers.

One of the earliest examples of ESL theories including non-classical effects is the First Order Shear Deformation Theory (FSDT) (Timoshenko, 1934, Mindlin, 1951 and Yang et al., 1966). However, Whitney and Pagano (1970) demonstrated that FSDT only yields improvements on CLA for global structural phenomenon but does not improve in-plane strain and stress predictions for highly heterogeneous and thick laminates. Furthermore, FSDT produces piecewise constant transverse shear stresses that violate IC and traction-free conditions at the top and bottom surfaces.

To overcome these shortcomings the so-called Higher-Order Shear Deformation Theories (HOT) were introduced. In general, the cross-section is allowed to deform in any form by including higher-order terms in the axiomatic expansions of the in-plane displacements u_x and u_y. Vlasov (1957) refined Mindlin's theory by guaranteeing that transverse shear strains and stresses disappear at the top and bottom surfaces in the absence of shear tractions. Taking Vlasov's condition into consideration, Reddy (1983)presented a higher-order shear deformation theory by expanding the in-plane displacement field to a third order polynomial in z. A large number of different shear shape functions have been published in the past ranging from polynomial (Ambartsumyan, 1958a, Reissner, 1975 and Reddy, 1986) to trigonometric (Levy, 1877,Stein, 1986, Touratier, 1991, Karama et al., 1998 and Ferreira et al., 2005), hyperbolic (Soldatos, 1992 and Neves et al., 2013) and exponential (Karama et al., 2003 and Mantari et al., 2011).

Transverse normal strains may be incorporated by extending the expansion of the out-of-plane displacement u_z to yield a class of theories denoted as Advanced Higher-Order Theories (AHOT). Here, the class of theory is often denoted by $\{a,b\}$ where a refers to the order of expansion of the in-plane displacements u_x and u_y, and b to the order of the transverse displacement u_z. Examples of such theories are given by Tessler, 1993,Cook and Tessler, 1998 and Barut et al., 2001 but these theories generally only provide improvements that are worth their additional computational effort for sandwich panels with compliant, thick cores (Demasi, 2012) or when one face laminate is considerably stiffer than the other (Gherlone, 2013).

All of the previously discussed theories are based on displacement formulations where the displacements u_x, u_y, u_z are treated as the unknown variables, and the strains and stresses are derived using kinematic and constitutive equations, respectively. In this case, the governing field and boundary equations may be derived using the principle of virtual displacements (PVD). Being formulated on a displacement-based assumption the transverse shear stresses typically do not guarantee the C_z^0-requirements *a priori*. More accurate transverse stresses are recovered *a posteriori* by integration of the in-plane stresses in Cauchy's 3D indefinite equilibrium equations (Whitney, 1972).

This post-processing operation can be precluded if some form of stress assumption is made. One class of model is based on applying the Hellinger–Reissner mixed variational principle. Here the strain energy is written in complementary form in terms of in-plane and transverse stresses, and the transverse equilibrium equation is introduced as a constraint condition using a Lagrange multiplier (Reissner, 1944 and Reissner, 1945).Batra and Vidoli, 2002 and Batra et al., 2002 used the Hellinger–Reissner mixed variational theorem to develop a higher-order theory for studying vibrations and plane waves in piezoelectric and anisotropic plates, accounting for both transverse shear and transverse normal deformations with all functional unknowns expanded in the thickness direction using orthonormal Legendre polynomials. The researchers showed that the major advantage of the Hellinger–Reissner theory is that by enforcing stresses to satisfy the natural boundary conditions at the top and bottom surfaces, and deriving transverse stresses from the plate equations directly, the stress fields are closer to 3D elasticity solutions than a pure displacement-based equivalent that relies on Hooke's law to derive the stress fields. In particular this means that boundary layers near clamped and free edges, and asymmetric stress profiles due to surface tractions on one surface only can be captured accurately. Cosentino and Weaver (2010) applied the Hellinger–Reissner principle to symmetrically laminated straight-fibre composites to develop a single sixth-order differential equation in just two variables: transverse deflection w and stress function Ω. The formulation of this theory is less general than the one proposed by Batra and Vidoli (2002) as its aims are to realise accurate 3D dimensional stress predictions for practical composite laminates at minimum computational cost.

Later, Reissner (1984) had the insight that when considering multi-layered structures, it is sufficient to restrict the stress assumptions to the transverse stresses because only these have to be specified independently to guarantee the IC requirements. This variational statement is known as Reissner's Mixed Variational Theory (RMVT), which makes model assumptions on the three displacements $u_x,\ u_y,\ u_z$ and independent assumptions on the transverse stresses $\tau_{xz},\ \tau_{yz},\ \sigma_{zz}$. Compatibility of the transverse strains derived from kinematic and constitutive equations is enforced by means of Lagrange multipliers.

HOT and AHOT provide considerable improvements in terms of transverse stress profiles and accurate modelling of global structural effects. However, these theories are not capable of explicitly capturing ZZ effects as the in-plane variables $u_x,\ u_y$ are defined to be at least C_z^1-continuous. In this regard ESL theories that incorporate ZZ kinematics present a good compromise between local, layerwise accuracy and computational cost. Based on an historical review of the topic by Carrera (2003a) the ZZ theories can generally be divided into three groups:

1. Lekhnitskii Multilayered Theory (LMT).
2. Ambartsumyan Multilayered Theory (AMT).
3. Reissner Multilayered Theory (RMT).

Lekhnitskii (1935) appears to be the first author to propose a ZZ theory originally formulated for multilayered beams. This was later extended to the analysis of plates byRen, 1986a and Ren, 1986b. Ambartsumyan, 1958a and Ambartsumyan, 1958bdeveloped a ZZ-theory for symmetric, specially orthotropic laminates by making a parabolic assumption for the transverse shear stresses. The corresponding displacement field is derived by integrating the shear strain kinematic equation through the thickness and solving for the integration constants by enforcing displacement-IC.Whitney (1969) later extended the analysis to symmetric laminates with off-axis plies and noted that the theory provides excellent results for gross laminate behaviour when compared to the 3D elasticity solutions of Pagano, 1969, Pagano, 1970a and Pagano, 1970b. However, Whitney also noticed that the theory fails to accurately capture the slope discontinuity at layer interfaces for the transverse shear stresses.

Di Sciuva, 1984 and Di Sciuva, 1985 introduced a displacement based ZZ theory where piece-wise linear, ZZ contributions in the thickness direction enhance a FSDT expansion for u_x and u_y. The slopes of the layer wise ZZ functions are obtained by enforcing the same transverse shear stress for all layers and by defining the ZZ function to vanish across the bottom layer. As a result, the transverse shear stress in all layers is identical to that of the bottom layer, causing a bias towards the transverse shear stiffness of this layer. To overcome this counterintuitive property Averill (1994) and Averill and Yip (1996) introduced a penalty term in the variational principle that enforces continuity of the transverse shear stresses as the penalty term becomes large. Tessler et al. (2007) note that the formulations based on Di Sciuva's early works present two major issues:

1. The in-plane strains are functions of the second derivative of transverse deflection w . This fact means less attractive C^1 continuous shape functions of w are required for implementation in FE codes.
2. The physical shear forces derived from the first derivatives of the bending moments are different from the shear forces derived by integrating the transverse shear stresses over the cross-section.

To remedy these drawbacks Tessler et al. developed a refined zig-zag theory (RZT) (Tessler et al., 2007, Tessler et al., 2009, Tessler et al., 2010a and Tessler et al., 2010b). The kinematics of RZT are essentially those of FSDT enhanced by a zig-zag field $\psi_\alpha(x,y)$ multiplied by a piecewise continuous transverse function ϕ_α^k,

$$u_\alpha^{(k)}(x,y,z) = u_\alpha + z\theta_\alpha + \phi_\alpha^{(k)}(z)\psi_\alpha \quad \text{for } \alpha = x,y. \tag{1a}$$
$$u_z(x,y,z) = w(x,y). \tag{1b}$$

In this theory the ZZ slopes $\beta_x = \partial\phi_x^{(k)}/\partial z$ and $\beta_y = \partial\phi_y^{(k)}/\partial z$ for u_x and u_y, respectively, are defined by the difference between the transverse shear rigidity of a layer $G_{\alpha z}$, and the effective transverse shear rigidity G of the entire layup

$$\beta_\alpha^k = \frac{G_{\alpha z}^k}{G_\alpha} - 1, \quad \text{and} \quad G_\alpha = \left[\frac{1}{t} \sum_{k=1}^{N} \frac{t^k}{G_{\alpha z}} \right]^{-1}, \tag{2}$$

where t^k and t are the thickness of the layer k and total laminate thickness, respectively. RZT has shown excellent results compared to the 3D elasticity solutions by Pagano, 1969 and Pagano, 1970a for both general composite laminations and sandwich constructions. Recently RZT has also been expanded to include transverse normal stretching and higher-order displacements for a ZZ theory of order {2, 2} (Barut et al., 2012).

Similarly, Murakami (1986) enhanced the axiomatic FSDT expansion by including a zig-zag function, herein denoted as Murakami's ZZ function (MZZF), that alternatively takes the values of +1 or −1 at layer interfaces. Therefore the slope purely depends on geometric differences between plies and is not based on continuity of transverse shear stresses. In addition Murakami made independent, piecewise parabolic assumptions for the transverse shear stresses and applied RMVT to obtain the governing field equations. In recent years the MZZF has been applied to functionally graded materials (Neves et al., 2013), sandwich structures (Brischetto et al., 2009a and Brischetto et al., 2009b) and in the framework of the Carrera Unified Formulation (Carrera et al., 2013a and Carrera et al., 2013b) for static and dynamic analyses. Carrera (2004) investigated the effect of including the MZZF in first-order and higher-order displacement-based and mixed-variational theories, showing that superior representation of displacements and stresses, combined with less computational cost can be achieved by including a single ZZ compared to a higher-order continuous term. On the other hand, Gherlone (2013)showed that the MZZF leads to inferior results than RZT for sandwiches with large face-to-core stiffness ratios and laminates with general layups. Thus, an accurate choice of the ZZ function seems to be of paramount importance.

A multiscale approach for modelling the multifaceted structural behaviour of composite laminates in one unified model has been proposed by Williams (2005). The theory uses a general framework with non-linear von Kármán displacement fields and additional temperature and solute diffusion variables on two lengthscales, global and local levels, with the transverse basis functions of the two lengthscales enforced to be independent. This results in two sets of variationally consistent governing equations such that the theory is capable of capturing, in a coupled fashion, the thermo-mechanical-diffusional phenomena of laminates at the micro, meso and macro levels simultaneously. The use of interfacial constitutive models allows the theory to model delamination initiation and growth, as well as non-linear elastic or inelastic interfacial constitutive relations in a unified form. Williams (2001) has shown that multi-lengthscale theories can be more computationally efficient than pure layerwise models as the order of theory can be increased on both the local and global level. The displacement-based theory features $3(V_g+V_1N)$ unknowns where V_g and V_1 are the global and local number of variables and N the number of layers. In general $V_g=V_1=3$ is sufficient for accurate 3D stress field predictions as derived from constitutive relations. In later work Williams (2008) developed an improved formulation by deriving the governing equations from the method of moments over different length scales and enforcing the interfacial continuity of transverse stresses in a strong sense.

The aim of this paper is to provide additional insight into the fundamental mechanics of the ZZ phenomenon and develop a computationally efficient analysis tool for industrial applications. The aim is not to develop a unified general theory as presented by Carrera, 2003b, Demasi, 2008, Batra and Vidoli, 2002 and Williams, 2005 but to provide evidence that non-classical effects due to highly heterogeneous laminations and their associated 3D stress fields can be captured adequately and efficiently by global third-order moments combined with a local ZZ moment. Section 2 outlines a physical explanation for the source of ZZ-displacement fields based on an analogous system of mechanical springs, which to the authors' knowledge, has previously not been given in this manner. In Section 3 these insights are used to combine Reddy's transverse shear function with an Ambartsumyan-type ZZ theory to derive a displacement-based model (MRZZ). In Section 4, the Hellinger–Reissner mixed variational principle is used to derive a ZZ theory that may be used alongside the RZT ZZ function (HR-RZT) or the MZZF (HR-MZZF). The theory is different from general theories in that in-plane and transverse stress fields share the same variables thereby greatly reducing the number of unknowns. In Section 5 the MRZZ, HR-RZT and HR-MZZF theories are compared to Pagano's exact 3D elasticity solutions (Pagano, 1969) in a simply-supported bending load case of a thick beam. Furthermore, the performance of the Hellinger–Reissner principle and the Reissner Mixed Variational Theory are compared. Finally, the influence of transverse shear, transverse normal and ZZ effects on bending deformations are analysed.

MECHANICS OF ZIG-ZAG DISPLACEMENTS

Origin of Zig-zag Displacements

In the following discussion our analysis is restricted to 1D composite beams with negligible transverse normal strain and stress effects. Thus, consider an N-layer composite beam of arbitrary constitutive properties as depicted in Fig. 1. The beam may be of entirely anisotropic or sandwich construction and is subjected to bending moments or shear forces causing it to deflect transversely to the stacking direction. In all cases the x-direction is defined to be along the principle beam axis (0° fibre-direction) while the z -axis is in the transverse stacking direction. Individual layers are prevented from sliding such that the IC conditions for the displacement field u_x and the transverse shear stress τxz are satisfied:

$$u_x^{(k)}(z_k) = u_x^{(k+1)}(z_k) \quad \text{and} \tag{3a}$$
$$\tau_{xz}^{(k)}(z_k) = \tau_{xz}^{(k+1)}(z_k), \quad k = 1 \ldots N-1, \tag{3b}$$

where superscripts and subscripts (k) indicate layerwise and interfacial quantities, respectively. If the composite beam comprises layers with different transverse shear modulii then the IC condition on transverse stress inherently results in discontinuous transverse shear strains across ply interfaces. Assuming linear geometric deformation, the kinematic relation for the transverse shear strain is given by

$$\gamma_{xz} = u_{z,x} + u_{x,z} \Rightarrow u_{x,z} = \gamma_{xz} - u_{z,x}, \tag{4}$$

where the comma notation denotes differentiation. As transverse normal strain is assumed to be negligible, i.e. u_z is constant for all layers, discontinuous transverse shear strains require a change in $u_{x,z}$ across ply interfaces. Thus the slope of the displacement field u_x in the thickness direction must change at ply interfaces giving rise to the so-called "zig-zag" displacement field. This effect is depicted graphically by the in-plane displacement (u_x), and transverse shear stress (τxz) and strain (γxz) plots through the thickness of a [90/0/90/0/90] laminate in Figs. 2(a) and (b), respectively. Here, the transverse shear modulus Gxz of the 90° layers are 2.5 times less than the value of the 0° oriented layers causing a step change in transverse shear strain at the ply interfaces.

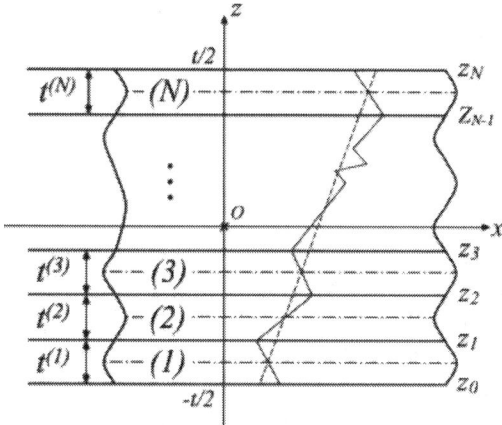

Figure 1. Arbitrary laminate configuration with co-ordinate system and approximate in-plane displacements.

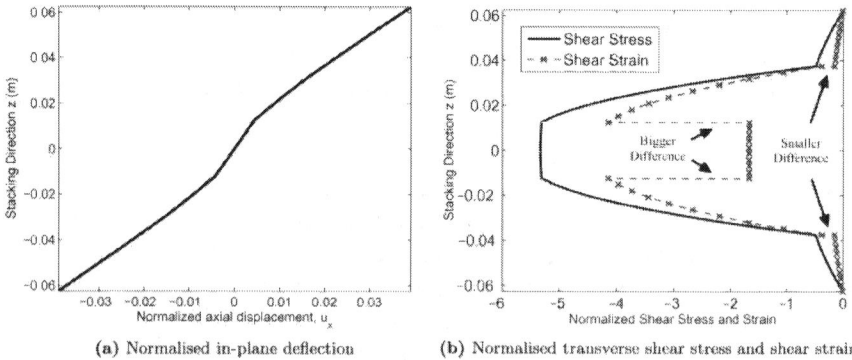

(a) Normalised in-plane deflection   **(b)** Normalised transverse shear stress and shear strain

Figure 2. Pagano's through-thickness solution of normalised in-plane deflection and transverse shear stress for a[90/0/90/0/90] laminate. It is an example of EWL indicated by the lack of zig-zag discontinuity at the outermost ply interfaces.

Fig. 2 also shows an example of a laminate with "Externally Weak Layers" (EWL). As discussed by Gherlone (2013) these laminates have external layers (k=1 or k=N) with transverse shear modulii lower than the adjacent internal layers (k=2 and k=N-1, respectively), and do not appear to have a ZZ displacement at these interfaces. Gherlone attributed this phenomenon to the stiffer inner layers dominating the more compliant external layers. Furthermore, these phenomena cannot be captured by RZT such that the layer material properties need to be altered artificially as follows:

EWL Implementation in RZT.

- If $G_{xz}^{(1)} < G_{xz}^{(2)}$, then $G_{xz}^{(1)} = G_{xz}^{(2)}$.

- If $G_{xz}^{(N)} < G_{xz}^{(N-1)}$, then $G_{xz}^{(N)} = G_{xz}^{(N-1)}$.

However, there is, in fact, a slope discontinuity at the interfaces of the EWL. This discontinuity is considerably less pronounced than at the interface between the internal 0° and 90° layers and is consequently not noticed as easily. This phenomenon may be explained by observing the general shape of the transverse shear stress and shear strain profile of a [90/0/90/0/90] laminate as shown in Fig. 2(b). The transverse shear stress at the interface between the outer layers is an order of magnitude less than the transverse shear stress at the inner interfaces. Therefore the discontinuity in transverse shear strain is much larger for inner layers than for outer layers, making it appear that there is no ZZ effect for the EWL. Even though the ratio of shear strains at the outer and inner [90/0] interface remains the same, the *difference* in magnitude is considerably larger for the inner layers. It is this *difference* in transverse shear strains, rather than the *ratio* that drives the slope discontinuity of the displacement field.

This also means that discontinuities in transverse shear strain for EWL laminates such as [0/90], [90/0] and [90/0/90] remain significant because of the larger transverse shear stresses at the EWL interfaces. Thus, Gherlone (2013) was forced to specify an exception to the EWL implementation rule required for RZT. The rule does not apply if the condition reduces the laminate to have the same transverse shear modulii for all layers, as would be the case for the [0/90], [90/0] and [90/0/90] laminates.

The difficulty in accurately modelling the ZZ phenomenon is that the displacement and transverse shear stress fields are interdependent. As shown in Fig. 2 the layerwise slopes of the ZZ-displacement field u_x depend on the transverse shear stress distribution. At the same time the transverse shear stress is a function of the kinematic equations. The difficulty in axiomatic, displacement-based theories is that the ad hoc assumptions for u_x and u_z need to derive accurate transverse shear stresses, if the IC on τxz is to be used to define layerwise ZZ slopes. Similarly, Ambartsumyan-type models (Ambartsumyan, 1958a) need to include all pertinent variables that influence the distribution of the transverse shear stress to derive an accurate through-thickness distribution for u_x.

Spring Model for Zig-zag Displacements

The IC requirements on in-plane displacements and transverse shear stresses are mechanically similar to a combined system of "springs-in-series" and "springs-in-parallel" (see Fig. 3). For example, a set of springs in series acted upon by a constant force extends the springs by different amounts. By analogy, a constant transverse shear stress acting on a laminate with layers of different shear modulii results in different shear strains in the layers. This represents a smeared, average value of the actual piecewise, parabolic transverse shear distribution. At the same time a system of springs in parallel elongated by a common displacement develops different reaction forces in the springs. This case may be interpreted as layerwise transverse shear stresses following the path of highest stiffness. Conceptually, these two spring systems combine to capture the interplay between transverse shear stress and strain as influenced by the IC conditions.

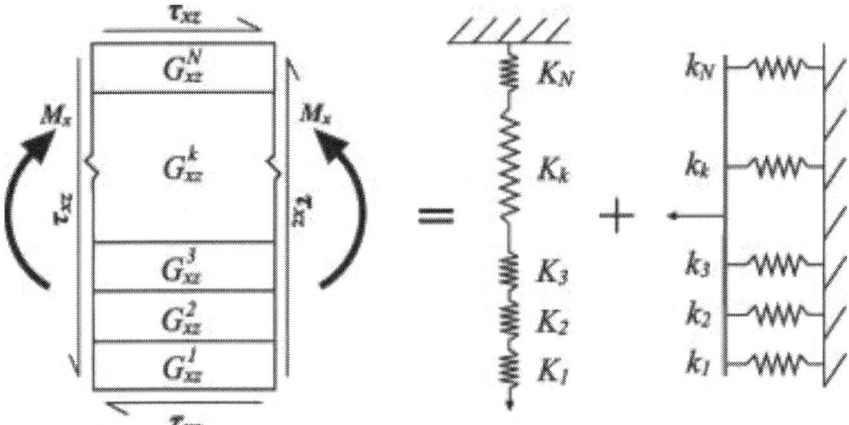

Figure 3. Schematic diagram of a composite laminate with varying layerwise transverse shear modulii $G_{xz}^{(k)}$ acted upon by transverse shear load and bending moment, which is modelled by an analogous system of mechanical springs.

The average transverse shear stress condition of the "springs-in-series" model is expressed via Hooke's law as an effective shear modulus G multiplied by an average shear strain $\bar{\gamma}_{xz}$

$$\tau_{xz} = G\bar{\gamma}_{xz}. \tag{5}$$

The effective shear modulus G is found using the stiffness equation of a set of springs in series

$$\frac{1}{K} = \frac{1}{k_1} + \frac{1}{k_2} + \ldots + \frac{1}{k_N},$$

$$G = \left[\frac{t^{(1)}/t}{G_{xz}^{(1)}} + \frac{t^{(2)}/t}{G_{xz}^{(2)}} + \ldots + \frac{t^{(N)}/t}{G_{xz}^{(N)}} \right]^{-1},$$

$$\therefore G = \left[\frac{1}{t} \sum_{k=1}^{N} \frac{t^{(k)}}{G_{..}^{(k)}} \right]^{-1}. \tag{6}$$

Note that the shear modulus of each layer is normalised by the layer thickness fraction to guarantee that $G=Gxz$ for a laminate with layers of equal shear modulii. It is worth emphasising that the effective stiffness G is the same as the expression for G_α in Eq.(2) found by Tessler et al. in RZT. The change in displacement slope $u_{x,z}$ at layer interfaces depends on the differences in transverse shear strain at interfaces. By inserting Eq. (5) into the transverse shear constitutive equation,

$$\gamma_{xz}^{(k)} = \frac{\tau_{xz}^{(k)}}{G_{xz}^k} = \frac{G}{G_{xz}^k} \bar{\gamma}_{xz} = g^{(k)} \bar{\gamma}_{xz}, \tag{7}$$

we see that the transverse shear strain is a function of the layerwise stiffness ratio $g^{(k)} = G/G_{xz}^{(k)}$. This ratio is used to capture the differences in layerwise displacement slopes.

Fig. 2(b) shows that the shear stress profile of a multi-layered beam differs from a single layer beam in that the z-direction curvature of the transverse shear stress profile in the stiffer $0°$ plies is increased whereas the curvature in the more compliant $90°$ is reduced. Integrating the axial stress σ_x derived from CLA for a zero B-matrix laminate in Cauchy's equilibrium equations,

$$\sigma_x = \bar{Q}^{(k)} \epsilon_x = \bar{Q}^{(k)} z \kappa_x,$$

$$\tau_{xz} = - \int \frac{d\sigma_x}{dx} dz = -\bar{Q}^{(k)} \frac{z^2}{2} \kappa_x + C$$

shows that the magnitude of the quadratic term z^2 is influenced by the transformed layer stiffness $\bar{Q}^{(k)}$, noting that κ_x is the flexural curvature. The "springs-in-series" analogy is now used to define an effective in-plane stiffness E,

$$E = \frac{1}{t} \sum_{k=1}^{N} t^{(k)} \bar{Q}^{(k)}. \tag{8}$$

The change in layerwise z-direction curvature of the transverse shear stress profile is a function of the relative magnitude of $\bar{Q}^{(k)}$ to the equivalent laminate stiffness. Therefore, a layerwise in-plane stiffness ratio $e^{(k)} = \bar{Q}^{(k)}/E$ is defined to quantify the change in transverse shear stress curvature of each layer.

MODIFIED REDDY ZIG-ZAG THEORY

In this section the laminate stiffness ratios $g^{(k)}$ and $e^{(k)}$ are used to derive a new Ambartsumyan-type ZZ theory by modifying Reddy's polynomial shear function (Reddy, 1983) to account for the ZZ effect. The use of the shear function guarantees that transverse shear stresses disappear at the top and bottom surfaces and that transverse shear stresses are continuous at layer interfaces. The model also resolves the issue of modelling "Externally Weak Layers" that was addressed by Gherlone (2013) and provides accurate *a priori* transverse shear predictions for composite laminates with zero B-matrix terms. Being a displacement-based method, the governing equations are derived using the principle of virtual displacements.

3.1. Transverse shear stress and displacement fields

As a starting point Hooke's Law of Eq. (5) is enhanced by the continuous, parabolic shear stress profile proposed by Reddy (1983),

$$\tau_{xz} = G\left(1 - \frac{4}{t^2}z^2\right)\bar{\gamma}_{xz}(x).$$

$$(9)$$

A numerical investigation of various composite and sandwich laminates using Pagano's 3D elasticity solution was performed. This showed that the layerwise z-direction curvatures of τxz can be quantified empirically by the modification factor $m^{(k)} = e^{(k)}(g^{(k)} + 1/g^{(k)} - 1)$. This modification factor reduces to $m^{(k)} = 1$ for a homogeneous laminate. Eq. (9) is rewritten to account for the differences in z-direction profile curvature for different layers,

$$\tau_{xz}^{(k)} = G\left\{A^{(k)} - \frac{4}{t^2}e^{(k)}\left(g^{(k)} + \frac{1}{g^{(k)}} - 1\right)z^2\right\}\bar{\gamma}_{xz}(x),$$

$$(10)$$

where the layerwise constants $A^{(k)}$ are found by enforcing transverse shear stress continuity at layer interfaces. Details of this derivation are shown in Appendix A.

From the constitutive relation the layerwise shear strain $\gamma_{xz}^{(k)}$ is given by

$$\gamma_{xz}^{(k)} = \frac{\tau_{xz}}{G_{xz}^{(k)}} = g^{(k)}\left\{A^{(k)} - \frac{4}{t^2}m^{(k)}z^2\right\}\bar{\gamma}_{xz}(x).$$
(11)

Next, the displacement field $u_x(x,z)$ is derived by integrating the kinematic equation(4) in the thickness z-direction while assuming that the transverse displacement $u_z(x,z)$ is constant through the thickness

$$u_x^{(k)}(x,z) = u_0(x) - zw_{,x}(x) + g^{(k)}\left\{A^{(k)}z - \frac{4}{3t^2}m^{(k)}z^3\right\}\bar{\gamma}_{xz} + c^{(k)}\bar{\gamma}_{xz},$$
(12a)

$$u_z(x,z) = w(x).$$
(12b)

The layerwise constants $c^{(k)}$ are found by enforcing continuity of displacements at layer interfaces and the condition that $u_x(x,0)=0$ due to the midplane symmetry of the beam. The derivation of the layerwise constants $c^{(k)}$ is provided in Appendix A.

Derivation of the Governing Equations

Consider a beam as represented in Fig. 4 undergoing static deformations under a specific set of externally applied loads and boundary conditions. The static behaviour of this structure is analysed using the ZZ displacement fields derived in Eq. (12a) and (12b)by means of the two kinematic unknowns $w(x)$ and $\bar{\gamma}_{xz}(x)$.

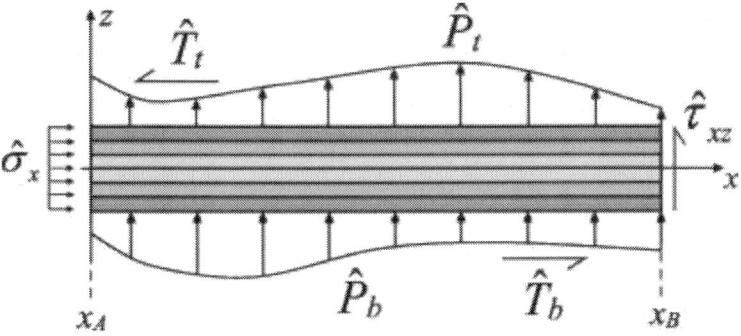

Figure 4. A composite beam loaded by distributed loads on the top and bottom surfaces and subjected to pertinent boundary conditions at ends A and B.

The principle of virtual displacements states that a body is in equilibrium if the virtual work done by the equilibrium forces, when the body is perturbed by a virtual amount $\delta\vec{u}$ from the true configuration $\vec{u}$ is zero. With regard to the elastic body depicted in Fig. 4 the virtual work done by the virtual displacement $\delta\vec{u}$ is

$$\delta\Pi_{PVD} = \int_{x_A}^{x_B} \int_S \left[\sigma_x^{H(k)} \delta\epsilon_x^{G(k)}(\vec{u}) + \tau_{xz}^H \delta\gamma_{xz}^{G(k)}(\vec{u}) \right] dSdx - \int_{x_A}^{x_B} q\delta w dx$$

$$- \int_{S_1} [\hat{\sigma}_x \delta u_x + \hat{\tau}_{xz} \delta w] dS_1,$$

where $\hat{\sigma}_x$ and $\hat{\tau}_{xz}$ are the prescribed stresses at the boundary points x_A and x_B. The superscript G indicates that the strains are calculated via the geometric strain–displacement relations:

$$\epsilon_x^{G(k)}(x,z) = u_{0,x}(x) - zw_{,xx}(x) + f^{(k)}(z)\bar{\gamma}_{xz,x}(x) \tag{13a}$$

$$\gamma_{xz}^{G(k)}(x,z) = g^{(k)}s^{(k)}(z)\bar{\gamma}_{xz}(x), \tag{13b}$$

where the shear function $s^{(k)}(z) = A^{(k)} - \frac{4}{t^2}m^{(k)}z^2$ and the displacement function $f^{(k)}(z) = g^{(k)}(A^{(k)}z - \frac{4}{3t^2}m^{(k)}z^3) + c^{(k)}$. The superscript H indicates that the stresses are calculated via the material constitutive equations (Hooke's Law)

$$\sigma_x^{H(k)} = \bar{Q}^{(k)}\epsilon_x^{G(k)}, \tag{14a}$$

$$\tau_{xz}^H = G_{xz}^{(k)}\gamma_{xz}^{G(k)} = Gs^{(k)}(z)\bar{\gamma}_{xz}, \tag{14b}$$

where $\bar{Q}^{(k)} = E_x^{(k)}$ for a beam in plane-stress in the width-direction, and $\bar{Q}^{(k)} = E_x^{(k)}/\left(1 - v_{xy}^{(k)}v_{yx}^{(k)}\right)$ for the plane-strain condition. Here $E_x^{(k)}$ and $G_{xz}^{(k)}$ are the Young's and transverse shear modulii, respectively while $v_{xy}^{(k)}$ and $v_{yx}^{(k)}$ are the major and minor Poisson's ratios of the k th layer, respectively. The energy functional Π is minimised by means of the calculus of variations to give two Euler–Lagrange equations that govern the equilibrium of force and moment resultants,

$$\delta w: \quad M_{x,xx} + q = 0, \tag{15a}$$

$$\delta\bar{\gamma}_{xz}: \quad L_{x,x} - Q_x = 0.. \tag{15b}$$

Furthermore, minimisation of Π gives the essential and natural boundary conditions at ends x_A and x_B:

$$\delta w: \quad M_{x,x} - \hat{V}_x = 0, \tag{16a}$$

$$\delta w_{,x}: \quad M_x - \hat{M}_x = 0, \tag{16b}$$

$$\delta \bar{\gamma}_{xz}: \quad L_x - \hat{L}_x = 0. \tag{16c}$$

The stress resultants are defined as follows:

$$\begin{pmatrix} M_x \\ L_x \end{pmatrix} = \int_{-t/2}^{t/2} \begin{pmatrix} z \\ f^{(k)} \end{pmatrix} \sigma_x^{H(k)} dz = \begin{bmatrix} D & D^{\gamma} \\ D^{\gamma} & D^{\gamma\gamma} \end{bmatrix} \begin{pmatrix} -w_{,xx} \\ \bar{\gamma}_{xz,x} \end{pmatrix}, \tag{17a}$$

$$Q_x = \int_{-t/2}^{t/2} g^{(k)} s^{(k)}(z) \tau_{xz}^{H(k)} dz = J\bar{\gamma}_{xz}, \tag{17b}$$

$$\hat{V}_x = \int_{-t/2}^{t/2} \hat{\tau}_{xz} dz, \tag{17c}$$

where the beam stiffness constants are calculated using the following integrals:

$$(D, D^{\gamma}, D^{\gamma\gamma}) = \sum_{k=1}^{N} \int_{z_{(k-1)}}^{z_{(k)}} \bar{Q}^{(k)} \left(z^2, z f^{(k)}, f^{(k)^2} \right) dz, \tag{18a}$$

$$J = \sum_{k=1}^{N} \int_{z_{(k-1)}}^{z_{(k)}} G g^{(k)} \left(s^{(k)}(z) \right)^2 dz. \tag{18b}$$

M_x is the bending moment as defined in CLA whereas the stress resultant L_x is a higher-order moment that captures the ZZ behaviour. Similarly, $\hat{M}_x$ and $\hat{L}_x$ are the prescribed moments on the boundary. Q_x represents the transverse shear force in the field equations whereas $\hat{V}_x$ is the shear force on the boundary. It is worth mentioning that Eq. (17b) and (17c) show an inconsistency between the two shear forces Q_x and $\hat{V}_x$ that is discussed by Groh and Weaver, 2015.

HELLINGER–REISSNER ZIG-ZAG THEORY

Previous studies (Tessler et al., 2007, Tessler et al., 2009 and Whitney, 1972) have shown that accurate transverse stress fields can be obtained in a post-processing step by integrating the axial stresses in Cauchy's indefinite equilibrium equations. It is expedient to perform this step *a priori* based on an accurate assumption of the axial stresses, and then derive new sets of governing equations using the Hellinger–Reissner mixed variational principle (Reissner, 1945). This approach was recently applied to straight-fibre and variable stiffness composites (Cosentino and Weaver, 2010 and Groh et al., 2013) but the authors did not include cubic axial stresses, ZZ effects and the possibility of modelling laminates with non-zero B-matrix terms in their models. The cubic behaviour is independent of the ZZ effect and is driven by the material orthotropy ratio E_x/Gxz and aspect ratio t/L that causes a "stress-channelling" effect towards the surfaces of the beam (Everstine and Pipkin, 1971 and Groh and Weaver, 2015). These previous works are extended here to remedy these shortcomings.

Higher-order Zig-zag Theory

We assume a cubic in-plane displacement field of the form,

$$u_x^{(k)}(x,z) = u_0 + z\theta + z^2\zeta + z^3\xi + \phi^{(k)}(z)\psi = \boldsymbol{f}_\phi^{(k)}(z) \cdot \mathcal{U}, \tag{19a}$$

$$u_z(x) = w, \tag{19b}$$

where u_0 is the reference surface axial displacement, θ is the rotation of the beam cross-section, ζ and ξ are higher-order rotations, ψ is the ZZ rotation and $\phi^{(k)}$ is a pertinent ZZ function. The row vector $\boldsymbol{f}_\phi^{(k)}$ describes the through-thickness displacement variation and U is the vector of in-plane variables,

$$\boldsymbol{f}_\phi^{(k)}(z) = \begin{bmatrix} 1 & z & z^2 & z^3 & \phi^{(k)}(z) \end{bmatrix}, \tag{20}$$

$$\mathcal{U} = \begin{bmatrix} u_0 & \theta & \zeta & \xi & \psi \end{bmatrix}^T. \tag{21}$$

As outlined in the work by Tessler et al. (2009) the RZT ZZ-function $\phi_{RZT}^{(k)}$ is defined by,

$$\phi_{RZT}^{(1)} = \left(z+\frac{t}{2}\right)\left(\frac{G}{G_{xz}^{(1)}}-1\right), \tag{22a}$$

$$\phi_{RZT}^{(k)} = \left(z+\frac{t}{2}\right)\left(\frac{G}{G_{xz}^{(k)}}-1\right) + \sum_{i=2}^{k} t^{(i-1)}\left(\frac{G}{G_{xz}^{(i-1)}}-\frac{G}{G_{xz}^{(k)}}\right), \tag{22b}$$

where G is the equivalent "springs-in-series" stiffness defined in Eq. (6). Murakami's ZZ function (MZZF) (Murakami, 1986) is given by

$$\phi^{(k)}_{MZZF} = (-1)^k \frac{2}{t^k} \left(z - z^k_m \right), \tag{23}$$

where z^k_m is the midplane co-ordinate of layer k. The layerwise definitions in Eqs.(22a), (22b) and (23) are re-written in the following general form

$$\phi^{(k)}(z) = g^{(k)}_{ZZF} \cdot z + c^{(k)}_{ZZF}, \tag{24}$$

where $g^{(k)}_{ZZF}$ and $c^{(k)}_{ZZF}$ are the z-coefficient and constant term of either the RZT or MZZF definitions. In Section 5 the accuracy of the RZT ZZ function and the MZZF are compared for a number of laminates and these two implementations are denoted by HR-RZT and HR-MZZF, respectively.

The following in-plane stress resultants can be derived from the displacement field of Eq.(19a) and (19b)

$$\mathcal{F} = \begin{bmatrix} N & M & O & P & L \end{bmatrix}^T = \int_{-t/2}^{t/2} f^{(k)^T}_\phi \sigma^{H(k)}_x dz,$$

$$\therefore \mathcal{F} = \begin{bmatrix} A & B & D & E & B^\phi \\ B & D & E & F & D^\phi \\ D & E & F & H & E^\phi \\ E & F & H & I & F^\phi \\ B^\phi & D^\phi & E^\phi & F^\phi & D^{\phi\phi} \end{bmatrix} \begin{pmatrix} u_{0,x} \\ \theta_{,x} \\ \zeta_{,x} \\ \xi_{,x} \\ \psi_{,x} \end{pmatrix} = S\mathcal{U}_x = S, \epsilon \tag{25}$$

where S is the constitutive stiffness matrix between stress resultants F and strains ϵ. The beam stiffness constants are calculated using the following integrals,

$$(A, B, D) = \sum_{k=1}^N \int_{z_{k-1}}^{z_k} \bar{Q}^{(k)} \left(1, z, z^2 \right) dz, \tag{26a}$$

$$(E, F, H, I) = \sum_{k=1}^N \int_{z_{k-1}}^{z_k} \bar{Q}^{(k)} \left(z^3, z^4, z^5, z^6 \right) dz, \tag{26b}$$

$$\left(B^\phi, D^\phi, D^{\phi\phi} \right) = \sum_{k=1}^N \int_{z_{k-1}}^{z_k} \bar{Q}^{(k)} \phi^{(k)} \left(1, z, \phi^{(k)} \right) dz, \tag{26c}$$

$$\left(E^\phi, F^\phi \right) = \sum_{k=1}^N \int_{z_{k-1}}^{z_k} \bar{Q}^{(k)} \phi^{(k)} \left(z^2, z^3 \right) dz. \tag{26d}$$

Here (O, P) and L are higher-order moments that capture the "stress-channelling" and ZZ effects, respectively. The axial stress field of this higher-order theory, written in terms of the stress resultants F, can be used alongside the Hellinger–Reissner mixed variational principle (HR) to develop a new higher-order ZZ theory that enforces Cauchy equilibrium equations *a priori*.

Derivation of Transverse Shear and Transverse Normal Stresses
The axial strain corresponding to the displacement field of Eq. (19a) and (19b) is given by,

$$\epsilon_x^{(k)} = \begin{bmatrix} 1 & z & z^2 & z^3 & \phi^{(k)}(z) \end{bmatrix} \cdot \mathcal{U}_x = \boldsymbol{f}_\phi^{(k)} \cdot \epsilon. \tag{27}$$

The constitutive relation between stress resultants F and strains ϵ in Eq. (25) is inverted to define a compliance matrix s

$$\epsilon = s\mathcal{F} \quad \text{where} \quad \boldsymbol{s} = \boldsymbol{S}^{-1}. \tag{28}$$

Applying the stress–strain Eq. (14a) in combination with Eqs. (27) and (28) the axial stress is,

$$\sigma_x^{(k)} = \bar{Q}^{(k)}\epsilon_x^{(k)} = \bar{Q}^{(k)}\boldsymbol{f}_\phi^{(k)}\boldsymbol{s}\mathcal{F}. \tag{29}$$

An expression for the transverse shear stress is found by integrating the axial stress of Eq. (29) in Cauchy's indefinite equilibrium equation,

$$\tau_{xz}^{(k)} = -\int \frac{d\sigma_x}{dx}\, dz = -\bar{Q}^{(k)}\left(\int \boldsymbol{f}_\phi^{(k)}\, dz\right)\boldsymbol{s}\mathcal{F}_{,x} = -\bar{Q}^{(k)}\boldsymbol{g}_\phi^{(k)}\boldsymbol{s}\mathcal{F}_{,x} + \boldsymbol{a}^{(k)}, \tag{30}$$

where $\boldsymbol{g}_\phi^{(k)}(z)$ captures the quartic variation of $\tau_{xz}^{(k)}$ through each ply k of the laminate,

$$\boldsymbol{g}_\phi^{(k)}(z) = \begin{bmatrix} z & \frac{z^2}{2} & \frac{z^3}{3} & \frac{z^4}{4} & g_{ZZF}^{(k)}\frac{z^2}{2} + c_{ZZF}^{(k)}z \end{bmatrix}. \tag{31}$$

The N layerwise constants $a^{(k)}$ are found by enforcing N-1 interfacial continuity conditions $\tau_{xz}^{(k)}(z_{k-1}) = \tau_{xz}^{(k-1)}(z_{k-1})$ and one of the prescribed surface tractions, i.e. either the bottom surface $\tau_{xz}^{(1)}(z_0) = \hat{T}_b$ or the top surface $\tau_{xz}^{(N)}(z_N) = \hat{T}_t$. Here we choose to enforce the bottom surface traction such that the layerwise integration constants a^k are found to be

$$a^{(k)} = \sum_{i=1}^{k} \left[\bar{Q}^{(i)} \mathbf{g}_\phi^{(i)}(z_{i-1}) - \bar{Q}^{(i-1)} \mathbf{g}_\phi^{(i-1)}(z_{i-1}) \right] s\mathcal{F}_x + \hat{T}_b,$$

$$a^{(k)} = \alpha^{(k)} s\mathcal{F}_x + \hat{T}_b, \tag{32}$$

where by definition $\bar{Q}^0 = 0$.

In the derivation of Eq. (32) the surface traction on the top surface is not enforced explicitly. However, this condition is automatically satisfied if equilibrium of the axial stress field Eq. (29) and transverse shear stress Eq. (30) is guaranteed. As we are dealing with an equivalent single layer the equilibrium equation is integrated through the thickness z-direction,

$$\int_{z_0}^{z_N} \sigma_{x,x} dz + \int_{z_0}^{z_N} \tau_{xz,z} dz = N_{,x} + \tau_{xz}^{(N)}(z_N) - \tau_{xz}^{(1)}(z_0) = 0. \tag{33}$$

An expression for $N_{,x}$ is easily derived from Eq. (29),

$$N_{,x} = \sum_{k=1}^{N} \left[\bar{Q}^{(k)} \mathbf{g}_\phi^{(k)}(z_k) - \bar{Q}^{(k)} \mathbf{g}_\phi^{(k)}(z_{k-1}) \right] s\mathcal{F}_{,x}. \tag{34}$$

Now the only undefined quantity in Eq. (33) is $\tau_{xz}^{(N)}(z_N)$ and an expression for this is sought using Eq. (30), (32) and (34)

$$\tau_{xz}^{(N)}(z_N) = \left(-\bar{Q}^{(N)} \mathbf{g}_\phi^{(N)}(z_N) + \alpha^{(N)} \right) s\mathcal{F}_{,x} + \hat{T}_b$$

$$= -\sum_{k=1}^{N} \left[\bar{Q}^{(k)} \mathbf{g}_\phi^{(k)}(z_k) - \bar{Q}^{(k)} \mathbf{g}_\phi^{(k)}(z_{k-1}) \right] s\mathcal{F}_{,x} + \hat{T}_b$$

$$\therefore \tau_{xz}^{(N)}(z_N) = -N_{,x} + \hat{T}_b \tag{35}$$

such that by substituting back into Eq. (33) we have

$$N_x + \left(-N_x + \hat{T}_b\right) - \tau_{xz}^{(1)}(z_0) = 0 \tag{36}$$

and as $\tau_{xz}^{(1)}(z_0) = \hat{T}_b$ the expression in Eq. (36) is satisfied. Thus, as long as Eq. (33) is enforced in the theory the top surface shear traction is automatically recovered.

Next, an expression for the transverse normal stress is derived in a similar fashion. Integrating Cauchy's transverse equilibrium equation yields

$$\sigma_z^{(k)} = -\int \frac{d\tau_{xz}}{dx} dz = \int \left(\bar{Q}^{(k)} g_\phi^{(k)} - \alpha^{(k)}\right) s\mathcal{F}_{,xx} dz - \hat{T}_{b,x} z$$
$$= \left(\bar{Q}^{(k)} h_\phi^{(k)} - \alpha^{(k)} z\right) s\mathcal{F}_{,xx} - \hat{T}_{b,x} z + b^{(k)}, \tag{37}$$

where the transverse normal matrix $h_\phi^{(k)}$ is given by

$$h_\phi^{(k)}(z) = \begin{bmatrix} \frac{z^2}{2} & \frac{z^3}{6} & \frac{z^4}{12} & \frac{z^5}{20} & g_{ZZF}^{(k)} \frac{z^3}{6} + c_{ZZF}^{(k)} \frac{z^2}{2} \end{bmatrix}. \tag{38}$$

The N layerwise constants $b^{(k)}$ are found by enforcing the N-1 continuity conditions $\sigma_z^{(k)}(z_{k-1}) = \sigma_z^{(k-1)}(z_{k-1})$ and one of the prescribed surface tractions, i.e. either the bottom surface $\sigma_z^{(1)}(z_0) = \hat{P}_b$ or the top surface $\sigma_z^{(N)}(z_N) = \hat{P}_t$. We again choose to enforce the bottom surface traction and then show that the surface traction at the top surface is recovered if equilibrium of the transverse shear stress Eq. (30) and transverse normal stress Eq. (37) is guaranteed. By enforcing the N-1 continuity conditions and $\sigma_z^{(1)}(z_0) = \hat{P}_b$,

$$b^{(k)} = \sum_{i=1}^{k} \left[\bar{Q}^{(i-1)} h_\phi^{(i-1)}(z_{i-1}) - \bar{Q}^{(i)} h_\phi^{(i)}(z_{i-1}) + \left(\alpha^{(i)} - \alpha^{(i-1)}\right) z_{i-1}\right] s\mathcal{F}_{,xx}$$
$$+ \hat{T}_{b,x} z_0 + \hat{P}_b,$$

$$b^{(k)} = \beta^{(k)} s\mathcal{F}_{,xx} + \hat{T}_{b,x} z_0 + \hat{P}_b, \tag{39}$$

where by definition $\bar{Q}^0 = \alpha^0 = 0$. Integrating the equilibrium equation through the thickness z-direction,

$$\int_{z_0}^{z_N} \tau_{xz,x} dz + \int_{z_0}^{z_N} \sigma_{zz} dz = Q_{,x} + \sigma_z^{(N)}(z_N) - \sigma_z^{(1)}(z_0) = 0,$$
(40)

where Q is the transverse shear force. An expression for $Q_{,x}$ is derived by integrating Eq. (30) and substituting Eq. (32) for $\boldsymbol{a}^{(k)}$,

$$Q_{,x} = \sum_{k=1}^{N} \left[\bar{Q}^{(k)} \left(\boldsymbol{h}_\phi^{(k)}(z_{k-1}) - \boldsymbol{h}_\phi^{(k)}(z_k) \right) + \boldsymbol{\alpha}^{(k)} t^{(k)} \right] \boldsymbol{s} \mathcal{F}_{,xx} + \sum_{k=1}^{N} \hat{T}_{b,x} t^{(k)},$$
(41)

where $t^{(k)}$ is the thickness of the kth layer. An expression for $\sigma_z^{(N)}(z_N)$ is defined using Eq. (37), (39) and (41)

$$\sigma_z^{(N)}(z_N) = \left(\bar{Q}^{(N)} \boldsymbol{h}_\phi^{(N)}(z_N) - \boldsymbol{\alpha}^{(N)} z_N + \boldsymbol{\beta}^{(N)} \right) \boldsymbol{s} \mathcal{F}_{,xx} - \hat{T}_{b,x}(z_N - z_0) + \hat{P}_b$$

$$= \sum_{k=1}^{N} \left[\bar{Q}^{(k)} \left(\boldsymbol{h}_\phi^{(k)}(z_k) - \boldsymbol{h}_\phi^{(k)}(z_{k-1}) \right) - \boldsymbol{\alpha}^{(k)} t^{(k)} \right] \boldsymbol{s} \mathcal{F}_{,xx}$$

$$- \sum_{k=1}^{N} \hat{T}_{b,x} t^{(k)} + \hat{P}_b \Rightarrow \sigma_z^{(N)}(z_N) = -Q_{,x} + \hat{P}_b$$
(42)

such that by substituting back into Eq. (40) we have

$$Q_{,x} + \left(-Q_{,x} + \hat{P}_b \right) - \sigma_z^{(1)}(z_0) = 0$$
(43)

and as $\sigma_z^{(1)}(z_0) = \hat{P}_b$ the expression in Eq. (43) is satisfied. Thus, as long as Eq. (40) is enforced in the theory the top surface pressure is automatically recovered. Finally, for conciseness, the layerwise coefficients in the equations for $\tau_{xz}^{(k)}$ and $\sigma_z^{(k)}$, Eqs. (30) and (37) respectively, are each combined conveniently into single layerwise vectors such that,

$$\tau_{xz}^{(k)} = \boldsymbol{c}^{(k)} \boldsymbol{s} \mathcal{F}_{,x} + \hat{T}_b,$$
(44a)

$$\sigma_z^{(k)} = \boldsymbol{e}^{(k)} \boldsymbol{s} \mathcal{F}_{,xx} - \hat{T}_{b,x}(z - z_0) + \hat{P}_b.$$
(44b)

Hellinger–Reissner mixed Variational Principle

A new set of equilibrium equations is derived by means of minimising the potential energy functional Π defined in Castigliano's Theorem of Least Work. In this case, Π is a functional of the stress resultants F that define the internal strain energy of the beam and the work done by external tractions. As shown in the previous section, equilibrium equations (33) and (40) should be satisfied to guarantee that the transverse stresses are recovered accurately. First, the transverse shear force Q is eliminated from Eq. (40)using the moment equilibrium condition,

$$\int_{z_0}^{z_N} z(\sigma_{xx} + \tau_{xz,z})dz = M_{,x} - Q + \left[z_N \hat{T}_t - z_0 \hat{T}_b\right] = 0,$$

$$\therefore Q = M_{,x} + \left[z_N \hat{T}_t - z_0 \hat{T}_b\right] \tag{45}$$

such that equilibrium equation (40) becomes

$$M_{,xx} + z_N \hat{T}_{t,x} - z_0 \hat{T}_{b,x} + \hat{P}_t - \hat{P}_b = 0. \tag{56}$$

For equilibrium of the system the first variation of the potential energy functional Π must vanish in such a manner that equilibrium equations (33) and (46) are satisfied over the whole beam domain $x \in [x_A, x_B]$. Following the rules of the calculus of variations this condition is enforced using Lagrange multipliers $\lambda_1(x)$ and $\lambda_2(x)$, respectively, and adding these to the variation of functional Π.

$$\delta\Pi = \delta\left[\frac{1}{2}\int_V (\sigma_x \epsilon_x + \tau_{xz}\gamma_{xz} + \sigma_z \epsilon_z)dV - \int_{S_2} (\sigma_x \hat{u}_x^{(k)} + \tau_{xz}\hat{w})dS_2 \right.$$
$$\left. + \int \lambda_1\left(N_{,x} + \hat{T}_t - \hat{T}_b\right)dx + \int \lambda_2\left(M_{,xx} + z_N \hat{T}_{t,x} - z_0 \hat{T}_{b,x} + \hat{P}_t - \hat{P}_b\right)dx\right] = 0, \tag{47}$$

where $\hat{u}_x^{(k)}$ and $\hat{w}$ are the displacements defined on the boundary curve S_2.

In Eq. (47) the quantities σ_x, τ_{xz} and σ_z are defined by Eqs. (29), (44a) and (44b), respectively. The transverse shear strain $\gamma_{xz}^{(k)}$ is defined using the constitutive equation(14b),

$$\gamma_{xz}^{(k)} = \frac{\tau_{xz}^{(k)}}{G_{xz}^{(k)}} = \frac{1}{G_{xz}^{(k)}}\left(\mathbf{c}^{(k)}\mathbf{s}\mathscr{F}_{,x} + \hat{T}_b\right). \tag{48}$$

The transverse normal strain $\epsilon_z^{(k)}$ is derived from Hooke's Law, written in terms of the full compliance matrix S_{ij} in a state of plane strain in y, as this is the condition assumed in Section 5. Thus,

$$\epsilon_z^{(k)} = R_{13}^{(k)}\sigma_x^{(k)} + R_{33}^{(k)}\sigma_z^{(k)}, \quad \text{where} \quad R_{ij} = S_{ij} - \frac{S_{i2}S_{j2}}{S_{22}}$$

$$= R_{13}^{(k)}\bar{Q}^{(k)}f_\phi^{(k)}\mathbf{s}\mathcal{F} + R_{33}^{(k)}\left(e^{(k)}\mathbf{s}\mathcal{F}_{,xx} - \hat{T}_{b,x}(z - z_0) + \hat{P}_b\right).$$

$$(49)$$

The new set of governing equations is derived by substituting all stress and strain expressions Eqs. (29), (44a), (44b), (48) and (49) into Eq. (47) and setting the first variation to zero. The corresponding Euler–Lagrange field equations in terms of the functional unknowns $\lambda_1(x)$, $\lambda_2(x)$ and F are,

$$\delta\lambda_1 : \; N_{,x} + \hat{T}_t - \hat{T}_b = 0, \tag{50a}$$

$$\delta\lambda_2 : \; M_{,xx} + z_N\hat{T}_{t,x} - z_0\hat{T}_{b,x} + \hat{P}_t - \hat{P}_b = 0, \tag{50b}$$

$$\delta\mathcal{F} : \; \mathbf{s}^T\mathcal{F} - (\eta - \omega)^T\mathcal{F}_{,xx} + \rho^T\mathcal{F}_{,xxxx} + \frac{\omega_p^T}{2}\hat{P}_b + \rho_p^T\hat{P}_{b,xx}$$
$$- \left(\chi + \frac{\omega_t}{2}\right)^T\hat{T}_{b,x} - \rho_t^T\hat{T}_{b,xxx} + \Lambda_{eq} = \mathbf{0}, \tag{50c}$$

where the superscript T denotes the transpose of a matrix. The pertinent essential and natural boundary conditions are given by,

$$\delta\mathcal{F} = 0 \quad \text{or} \quad \left(\eta - \frac{\omega}{2}\right)^T\mathcal{F}_{,x} - \rho^T\mathcal{F}_{,xxx} + \chi^T\hat{T}_b + \rho_t^T\hat{T}_{b,xx}$$
$$- \rho_p^T\hat{P}_{b,x} + \Lambda_{bc_1} = \hat{\mathcal{U}}, \tag{51a}$$

$$\delta\mathcal{F}_{,x} = 0 \quad \text{or} \quad \frac{\omega^T}{2}\mathcal{F} + \rho^T\mathcal{F}_{,xx} - \rho_t^T\hat{T}_{b,x} + \rho_p^T\hat{P}_b + \Lambda_{bc_2} = \hat{\mathcal{W}}. \tag{51b}$$

The governing equations related to δF are written in matrix notation, with each row defining a separate equation. In total there are seven field equations that can be solved by defining the ten boundary equations and four surface tractions. Eq. (50c) are enhanced versions of the CLA constitutive equation $M=Dw_{,xx}$, taking into account transverse shearing and transverse normal effects, and the influence of the higher-order moments. The members of matrix η are transverse shear correction factors, while members of ρ and ω are correction factors related to the transverse normal stresses and Poisson's effect of transverse normal stresses, respectively. The members of row vectors χ, ρ_t, ρ_p, ω_t and ω_p are correction factors enforcing transverse shear and transverse normal behaviour related to the surface tractions. Column vectors Λ only include the Lagrange multipliers λ_1, λ_2 and their derivatives. The full derivation of the governing equations including details of all coefficients are given in Appendix B. Finally, the expressions found in Eq. (51a) can be used to determine the deformation vector U of the reference surface, whereas the second row of Eq. (51b) can be used to find an expression for the bending deflection w.

RESULTS AND MODEL VALIDATION

The governing equations for the Modified Reddy zig-zag theory (MRZZ) and the Hellinger–Reissner zig-zag theory (HR-RZT and HR-MZZF) were derived for laminated beams in plane-strain because this allows the results to be compared against Pagano's 3D elasticity solution. However, both theories are not restricted to beams and may readily be extended to the analysis of laminated plates.

Consider a multilayered, laminated beam comprising N orthotropic composite layers as illustrated in Fig. 4 with the mid-plane and normal to the beam aligned with the cartesian x- and z -axes. The layers may be arranged in any general fashion with different ply thicknesses, material properties or material orientations. The beam is assumed to be simply-supported at the two ends $x_A=0$ and $x_B=a$ as shown in Fig. 5, and is considered to undergo static deformation in plane strain under the applied sinusoidal distributed load equally divided between the top and bottom surfaces $\hat{P}^b = -\hat{P}^t = q_0/2 \cdot \sin(\pi x/a)$. This boundary value problem is analysed using the governing equations derived in Eq.(15a) and (15b) for the Modified Reddy displacement-based theory and Eq. (50a),(50b) and (50c) for the Hellinger–Reissner mixed theory.

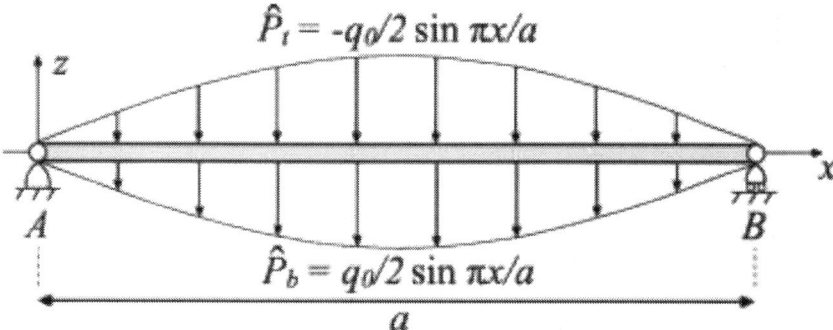

Figure 5. A simply supported beam loaded by a sinusoidal distributed load that is used to assess the accuracies of the presented ZZ formulations against the 3D elasticity solution of Pagano (1969).

Modified Reddy Zig-zag Theory

A closed form solution for $(W(x), \bar{\gamma}_{xz}(x))$ is sought that satisfies the boundary conditions at the two ends,

$$w = M_x = L_x = 0 \quad \text{for} \quad x = 0, a. \tag{52}$$

The expressions,

$$w = W \sin\left(\frac{\pi x}{a}\right), \tag{53a}$$

$$\bar{\gamma}_{xz} = \Gamma \cos\left(\frac{\pi x}{a}\right) \tag{53b}$$

satisfy the above boundary conditions exactly, where W and Γ are the bending deflection and shear rotation amplitudes, respectively. Substituting Eq. (53a) and (53b) into the constitutive relations (17a), (17b) and (17c) and then into the governing differential Eqs. (15a) and (15b) results in the following algebraic expressions for W and Γ,

$$\Gamma = -\frac{a^3}{\pi^3 \left(D^{\bar{\gamma}} - \frac{DD^{\bar{\gamma}\bar{\gamma}}}{D^{\bar{\gamma}}} - \frac{D^{\bar{\gamma}}_I}{D^{\bar{\gamma}}} \frac{a^2}{\pi^2}\right)} q_0, \tag{54a}$$

$$W = \left(\frac{D^{\bar{\gamma}\bar{\gamma}}}{D^{\bar{\gamma}}} \frac{a}{\pi} + \frac{J}{D^{\bar{\gamma}}} \frac{a^3}{\pi^3}\right) \Gamma. \tag{54b}$$

The coefficients (W, Γ) are now used to calculate displacements, strains and stresses using the pertinent formulae in Section 3.2.

Hellinger–Reissner Zig-zag Theory

For the Hellinger–Reissner zig-zag theories HR-RZT and HR-MZZF variable assumptions that satisfy the boundary conditions,

$$\hat{W} = \mathcal{F} = \mathbf{0} \quad \text{at} \quad x = 0, a. \tag{55}$$

are given by

$$(\lambda_2, \mathcal{F}) = (\lambda_{2_0}, \mathcal{F}_0) \cdot \sin\left(\frac{\pi x}{a}\right) \tag{56a}$$

$$\lambda_1 = \lambda_{1_0} \cdot \cos\left(\frac{\pi x}{a}\right) \tag{56b}$$

The boundary condition $N=0$ at $x=a$, b in Eq. (55) combined with the absence of surface shear tractions $\hat{T}_b = \hat{T}_t = 0$ means that the membrane force N vanishes over the whole beam domain. Therefore the membrane force amplitude $N_0 = 0$ in Eq. (56a) and equilibrium Eq. (50a) need not be considered.

Substituting Eq. (56a) and (56b) into the governing differential equations (50b) and (50c) results in six simultaneous algebraic equations in six unknowns $x = (M_0 \ O_0 \ P_0 \ L_0 \ \lambda_{1_0} \ \lambda_{2_0})^T$. These equations are readily solved by matrix inversion,

$$x = K^{-1}q, \tag{57}$$

where the stiffness matrix K is comprised of the coefficients of the $\mathcal{F}$, λ_1 and λ_2 terms in Eqs. (50b) and (50c) and the column load vector q is comprised of the terms associated with $\hat{T}_b, \hat{T}_t, \hat{P}_b$ and $\hat{T}_t$.

Numerical Results

The MRZZ and HR-RZT theories introduced within are compared against the 3D elasticity solution by Pagano (1969) for various symmetrically and arbitrarily laminated composite and sandwich beams. Even though the 3D elasticity solution by Pagano was developed for cylindrical bending of an infinitely wide plate, the solution is equally applicable to beams under plane strain. In order to emphasise the effects of transverse shear and ZZ deformability, relatively deep beams of length-to-thickness ratios $a/t = 8$ are considered for all stacking sequences. The material properties and stacking sequences are shown in Table 1 and Table 2, respectively. Material p was orginally defined by Pagano (1969) and is representative of a carbon-fibre reinforced plastic whereas material m features increased transverse stiffness and is based on the work byToledano and Murakami (1987). Material pvc is a closed-cell polyvinyl chloride foam modelled as an isotropic material. The honeycomb core h is modelled as transversely isotropic and features significantly lower transverse shear stiffness than material p to exacerbate the ZZ effect. As all results are normalised data, the Young's and shear modulii in Table 1 are presented in non-dimensionalised form, in this case with respect to the shear modulus G_{12} of material h.

Table 1. Mechanical properties of materials p, m, pvc and h normalised by the in-plane shear modulus $G_{12}^{(h)}$ of material h.

Material	$\dfrac{E_1}{G_{12}^{(h)}}$	$\dfrac{E_2}{G_{12}^{(h)}}$	$\dfrac{E_3}{G_{12}^{(h)}}$	v_{12}	v_{13}	v_{23}	$\dfrac{G_{12}}{G_{12}^{(h)}}$	$\dfrac{G_{13}}{G_{12}^{(h)}}$	$\dfrac{G_{23}}{G_{12}^{(h)}}$
p	25×10^6	1×10^6	1×10^6	0.25	0.25	0.25	5×10^5	5×10^5	2×10^5
m	32.57×10^6	1×10^6	10×10^6	0.25	0.25	0.25	6.5×10^5	8.21×10^6	3.28×10^6
pvc	25×10^4	25×10^4	25×10^4	0.3	0.3	0.3	9.62×10^4	9.62×10^4	9.62×10^4
h	250	250	2500	0.9	3×10^{-5}	3×10^{-5}	1	875	1750

Table 2. Analysed stacking sequences with zero B-matrix layups A–H and arbitrary layups I–M. Subscripts indicate the repetition of a property over the corresponding number of layers. Laminates with Externally Weak Layers are indicated by (EWL).

Laminate	Layer thickness ratio	Layer materials	Stacking sequence
A	$[(1/3)_3]$	$[p_3]$	$[0/90/0]$
B	$[.2_5]$	$[p_5]$	$[0/90/0/90/0]$
C (EWL)	$[.2_5]$	$[p_5]$	$[90/0/90/0/90]$
D	$[(1/51)_{51}]$	$[p_{51}]$	$[0/(90/0)_{25}]$
E	$[(1/30)_3/0.8/(1/30)_3]$	$[p_3/pvc/p_3]$	$[0/90/0_3/90/0]$
F	$[(1/30)_3/0.8/(1/30)_3]$	$[p_3/h/p_3]$	$[0/90/0_3/90/0]$
G (EWL)	$[.1_2/.2_3/.1_2]$	$[p_2/pvc/h/pvc/p_2]$	$[90/0_5/90]$
H (EWL)	$[(1/12)_{12}]$	$[p_{12}]$	$[\pm45/\mp45/0/90_2/0/\mp45/\pm45]$
I	$[0.3/0.7]$	$[p_2]$	$[0/90]$
J (EWL)	$[0.25_4]$	$[p_4]$	$[0/90/0/90]$
K	$[0.1/0.3/0.35/0.25]$	$[p_2/m/p]$	$[0/90/0_2]$
L (EWL)	$[0.3/0.2/0.15/0.25/0.1]$	$[p_3/m/p]$	$[0/90/0_2/90]$
M	$[0.1/0.7/0.2]$	$[m/pvc/p]$	$[0_3]$

The stacking sequences in Table 2 are split into a group of zero B-matrix laminates A-H and general laminates I-M. Laminates A-D are symmetric cross-ply composite laminates with $0°$ and $90°$ layers progressively more dispersed through the thickness. Even though thick blocks of $0°$ and $90°$ plies (as in Laminate A) are not commonly used in industry due to transverse cracking issues, this sequence maximises the ZZ effect for validation purposes. Laminates E-G are symmetric thick-core sandwich beams with uni-directional or cross-ply outer skins. Laminate G may be considered as a challenging test case in that the sandwich construction maximises the ZZ effect, the stacking sequence is a combination of three distinct materials and the $90°$ outer plies act as Externally Weak Layers (EWL). Laminate J is an example of an anti-symmetrically laminated beam with zero B-matrix terms. As Pagano's 3D elasticity solution does not include modelling of off-axis, anisotropic layers the $\pm 45°$ plies were modelled with effective orthotropic material properties using the transformed axial modulii $\bar{\bar{Q}}_x^{(k)}$ and transverse shear moduli $\bar{\bar{G}}_{xz}^{(k)}$. Laminates I and J are non-symmetric counterparts to the cross-ply laminates A-D mentioned above. Finally, laminates K-M are highly heterogeneous laminates with general laminations in terms of ply orientations, ply thicknesses and ply material properties, and present a challenging test case for any ESL theory.

Normalised quantities of the bending deflection w , axial stress σ_x, transverse shear stress τxz and transverse normal stress σ_z are used as metrics to assess the accuracy of the present theories. These normalised quantities are defined as follows,

$$\bar{w} = \frac{10^6 t^2}{q_0 a^4} \int_{-\frac{t}{2}}^{\frac{t}{2}} u_z\left(\frac{a}{2}, z\right) \, dz, \qquad \bar{\sigma}_x = \frac{t^2}{q_0 a^2} \cdot \sigma_x\left(\frac{a}{2}, z\right), \qquad \text{(58a-b)}$$

$$\bar{\tau}_{xz} = \frac{1}{q_0} \cdot \tau_{xz}(0, z), \qquad \bar{\sigma}_z = \frac{1}{q_0} \cdot \sigma_z\left(\frac{a}{2}, z\right) \qquad \text{(58c-d)}$$

and are calculated at the indicated locations (x, z) along the span of the beam. The normalised deflections $\bar{w}$ for the ESL theories are constant through the thickness and thus compared against Pagano's normalised average through-thickness deflection Eq.(58a).

The relative percentage errors with respect to Pagano's 3D elasticity solution of the normalised metrics $\bar{w}$, the maximum through-thickness values $\bar{\sigma}_x^{max}$ and $\bar{\tau}_{xz}^{max}$ for the zero B-matrix laminates A-H are shown in Table 3. In each case errors greater than 3% have been underlined to indicate an error outside the acceptable accuracy margin. For comparison, the table also includes the results of a third-order RMVT implementation using the cubic in-plane displacement assumption of Lo et al. (1977) enhanced with a ZZ variable, combined with the piecewise, parabolic transverse shear stress assumption ofMurakami (1986). Both the HR and RMVT formulations have been implemented with the RZT ZZ function and Murakami's ZZ function (MZZF) (Murakami, 1986) to compare their performances. Similarly, the results for the general laminates I-M are shown in Table 4. This table does not include the results for MRZZ as this theory is based on the assumption of zero B-matrix lamination. Finally, to qualitatively compare the stress fields through the full laminate thicknesses, the normalised axial stresses $\overline{\sigma}_x$ and transverse shear stresses $\overline{\tau}_{xz}$ are plotted in Fig. 6, Fig. 7, Fig. 8, Fig. 9, Fig. 10, Fig. 11, Fig. 12, Fig. 13, Fig. 14, Fig. 15, Fig. 16, Fig. 17 and Fig. 18.

Table 3. Zero B-matrix laminates A–H: Normalised results of maximum transverse deflection, maximum absolute axial stress and maximum absolute transverse shear stress of Pagano's solution are shown in bold. Different model results are given by percentage errors with respect to Pagano's solution. Errors greater than 3% are underlined.

Laminate	Model	$\overline{w}$	$\overline{\sigma}_x^{max}$	$\overline{\tau}_{xz}^{max}$
A	*Pagano*	*0.0116*	*0.7913*	*3.3167*
	MRZZ (%)	0.03	1.08	2.29
	HR-RZT (%)	0.06	−0.23	−0.04
	HR-MZZF (%)	0.05	−0.23	−0.04
	RMVT-RZT (%)	0.07	−2.03	0.55
	RMVT-MZZF (%)	0.07	−2.03	0.55
B	*Pagano*	*0.0124*	*0.8672*	*3.3228*
	MRZZ (%)	0.20	0.36	−1.63
	HR-RZT (%)	0.07	−0.92	−0.23
	HR-MZZF (%)	0.07	−0.92	−0.23
	RMVT-RZT (%)	0.08	−1.10	−1.40
	RMVT-MZZF (%)	0.08	−1.10	−1.40

C (EWL)	*Pagano*	*0.0303*	*1.6307*	*5.3340*
	MRZZ (%)	0.36	0.56	1.36
	HR-RZT (%)	0.24	−0.49	0.03
	HR-MZZF (%)	0.24	1.05	0.07
	RMVT-RZT (%)	−0.66	−0.48	0.37
	RMVT-MZZF (%)	−1.49	0.45	18.06
D	*Pagano*	*0.0154*	*1.2239*	*3.6523*
	MRZZ (%)	0.23	0.63	1.94
	HR-RZT (%)	0.11	0.34	−0.05
	HR-MZZF (%)	0.11	0.34	−0.05
	RMVT-RZT (%)	−0.62	−1.15	19.22
	RMVT-MZZF (%)	−0.62	−1.15	19.22
E	*Pagano*	*0.0309*	*1.9593*	*2.8329*
	MRZZ (%)	0.23	0.39	−0.69
	HR-RZT (%)	0.06	0.02	−0.16
	HR-MZZF (%)	0.09	−0.88	−0.33
	RMVT-RZT (%)	0.13	0.08	31.96
	RMVT-MZZF (%)	−1.18	−0.04	218.78
F	*Pagano*	*1.0645*	*13.9883*	*8.1112*
	MRZZ (%)	−0.70	0.28	−79.54
	HR-RZT (%)	−0.28	−0.24	0.05
	HR-MZZF (%)	−0.25	7.96	−0.29
	RMVT-RZT (%)	−0.32	−0.15	16.74

	RMVT-MZZF (%)	$\underline{-62.92}$	$\underline{-54.24}$	$\underline{2697.28}$
G (EWL)	*Pagano*	*0.4590*	*6.3417*	*5.6996*
	MRZZ (%)	−0.37	0.43	$\underline{-61.92}$
	HR-RZT (%)	−0.02	0.02	0.04
	HR-MZZF (%)	$\underline{7.11}$	$\underline{10.66}$	−0.13
	RMVT-RZT (%)	−0.08	0.07	$\underline{5.53}$
	RMVT-MZZF (%)	$\underline{-88.80}$	$\underline{-70.05}$	$\underline{188.93}$
H (EWL)	*Pagano*	*0.0224*	*0.6157*	*4.0096*
	MRZZ (%)	0.47	0.55	$\underline{4.13}$
	HR-RZT (%)	0.40	0.26	0.05
	HR-MZZF (%)	0.48	2.75	1.07
	RMVT-RZT (%)	0.45	−0.06	−0.26
	RMVT-MZZF (%)	$\underline{-3.15}$	−0.64	$\underline{42.11}$

Table 4. Arbitrary laminates I–M: Normalised results of maximum transverse deflection, maximum absolute axial stress and maximum absolute transverse shear stress of Pagano's solution are shown in bold. Different model results are given by percentage errors with respect to Pagano's solution. Errors greater than 3% are underlined.

Laminate	Model	$\overline{w}$	$\overline{\sigma}_x^{max}$	$\overline{\tau}_{xz}^{max}$
I	*Pagano*	*0.0482*	*2.0870*	*4.8799*
	HR-RZT (%)	0.64	−0.59	0.17
	HR-MZZF (%)	0.64	−0.59	0.17
	RMVT-RZT (%)	0.57	−1.84	0.41
	RMVT-MZZF (%)	0.57	−1.84	0.41
J (EWL)	*Pagano*	*0.0195*	*1.2175*	*4.3539*
	HR-RZT (%)	0.36	−0.94	0.06
	HR-MZZF (%)	0.36	0.67	0.10

	RMVT-RZT (%)	−0.39	−2.22	3.71
	RMVT-MZZF (%)	−0.81	−0.69	11.38
K	*Pagano*	*0.0100*	*0.9566*	*4.1235*
	HR-RZT (%)	0.39	−0.06	−0.48
	HR-MZZF (%)	0.39	0.19	0.11
	RMVT-RZT (%)	−5.48	−4.42	8.95
	RMVT-MZZF (%)	−0.67	−1.05	13.56
L (EWL)	*Pagano*	*0.0115*	*1.0368*	*3.8037*
	HR-RZT (%)	0.29	0.61	−0.12
	HR-MZZF (%)	0.53	6.16	0.17
	RMVT-RZT (%)	0.12	0.05	0.91
	RMVT-MZZF (%)	−12.48	−3.93	195.58
M	*Pagano*	*0.0226*	*1.4902*	*2.8969*
	HR-RZT (%)	0.05	0.51	−0.06
	HR-MZZF (%)	0.06	1.11	0.05
	RMVT-RZT (%)	0.03	0.47	−0.22
	RMVT-MZZF (%)	0.05	0.98	3.91

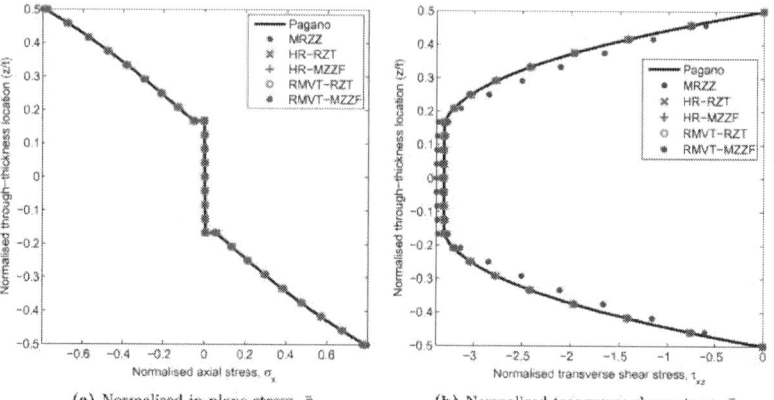

(a) Normalised in-plane stress, $\bar{\sigma}_x$ (b) Normalised transverse shear stress, $\bar{\tau}_{zx}$

Figure 6. Laminate A: through-thickness distribution of the normalised in-plane stress and transverse shear stress.

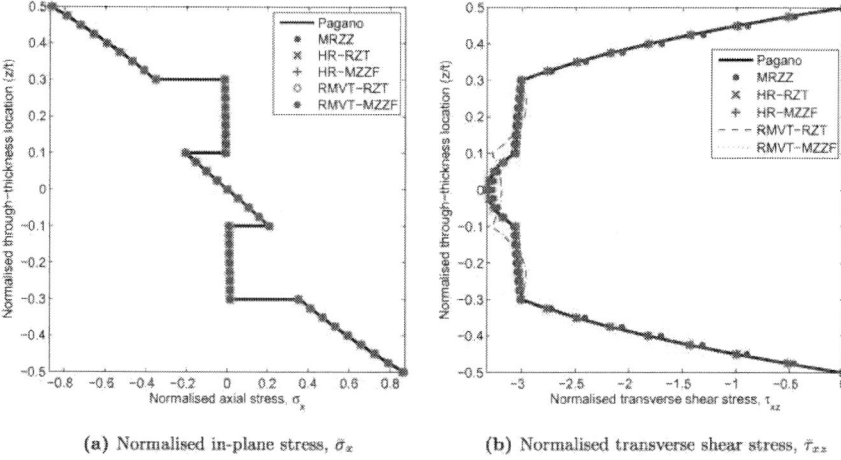

(a) Normalised in-plane stress, $\bar{\sigma}_x$

(b) Normalised transverse shear stress, $\bar{\tau}_{xz}$

Figure 7. Laminate B: through-thickness distribution of the normalised in-plane stress and transverse shear stress.

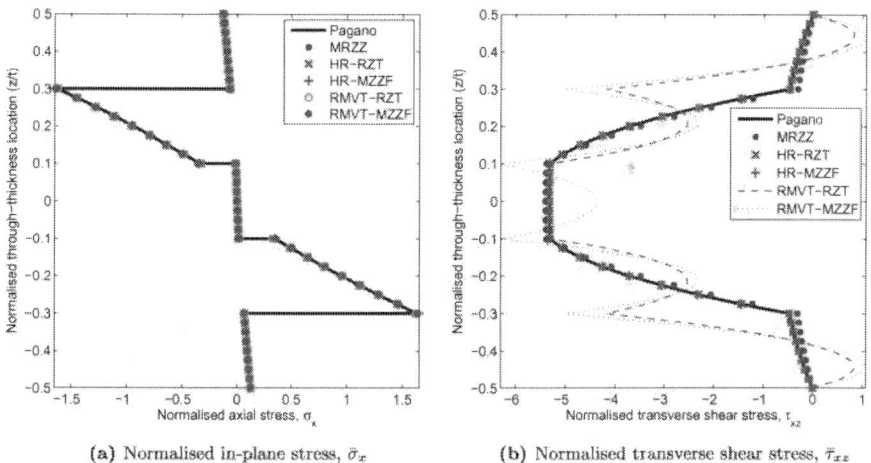

(a) Normalised in-plane stress, $\bar{\sigma}_x$

(b) Normalised transverse shear stress, $\bar{\tau}_{xz}$

Figure 8. Laminate C: through-thickness distribution of the normalised in-plane stress and transverse shear stress.

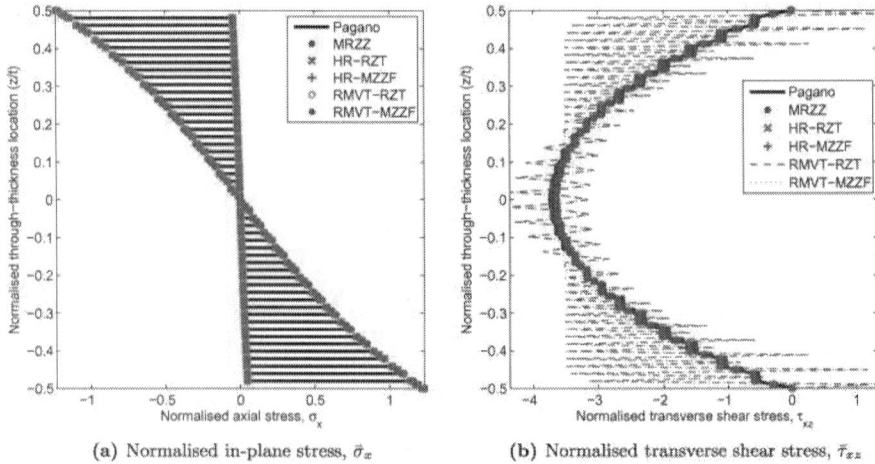

(a) Normalised in-plane stress, $\bar{\sigma}_x$

(b) Normalised transverse shear stress, $\bar{\tau}_{xz}$

Figure 9. Laminate D: through-thickness distribution of the normalised in-plane stress and transverse shear stress.

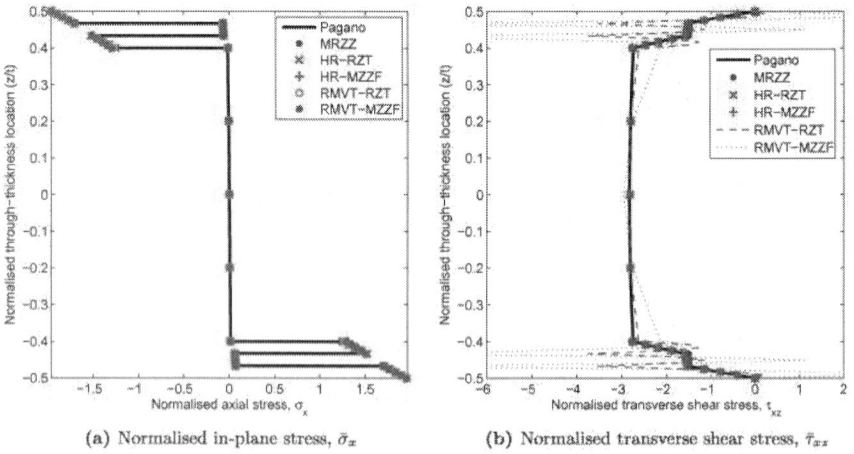

(a) Normalised in-plane stress, $\bar{\sigma}_x$

(b) Normalised transverse shear stress, $\bar{\tau}_{xz}$

Figure 10. Laminate E: through-thickness distribution of the normalised in-plane stress and transverse shear stress.

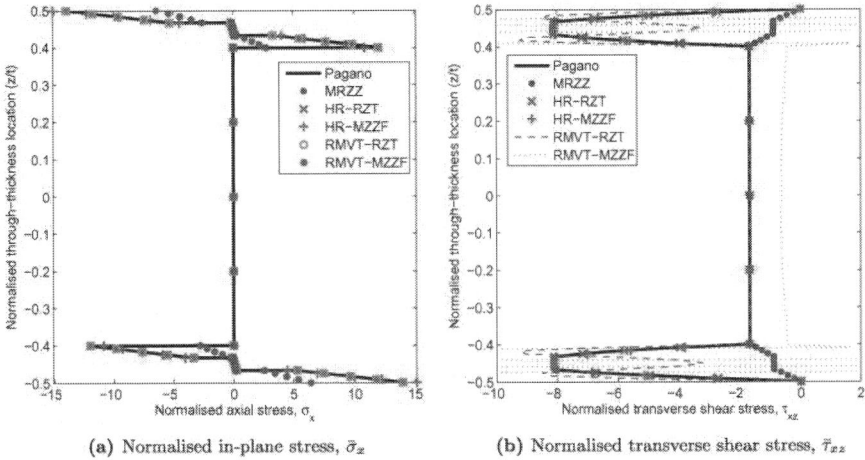

(a) Normalised in-plane stress, $\bar{\sigma}_x$

(b) Normalised transverse shear stress, $\bar{\tau}_{xz}$

Figure 11. Laminate F: through-thickness distribution of the normalised in-plane stress and transverse shear stress.

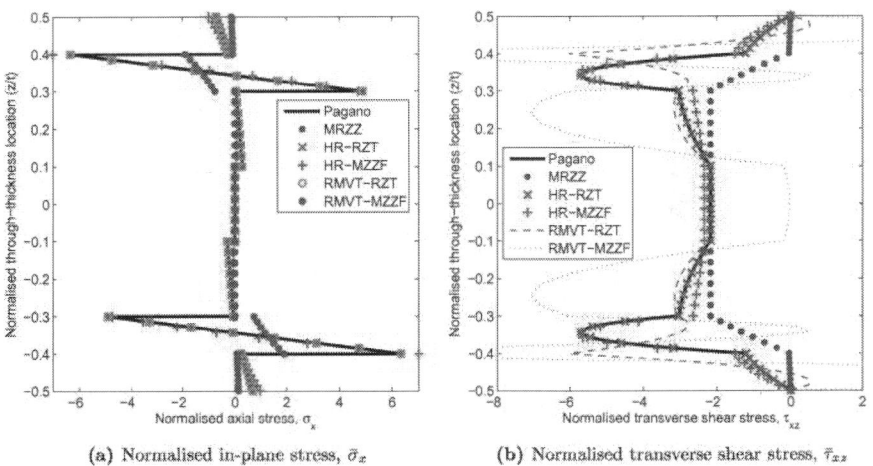

(a) Normalised in-plane stress, $\bar{\sigma}_x$

(b) Normalised transverse shear stress, $\bar{\tau}_{xz}$

Figure 12. Laminate G: through-thickness distribution of the normalised in-plane stress and transverse shear stress.

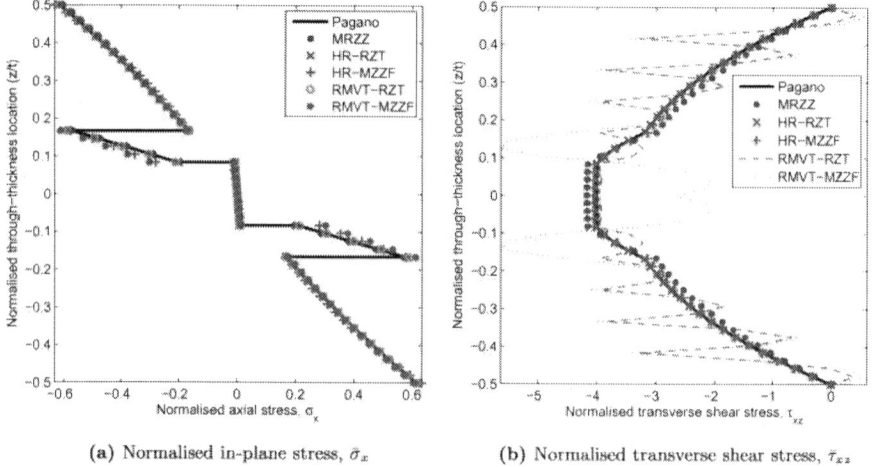

(a) Normalised in-plane stress, $\bar{\sigma}_x$

(b) Normalised transverse shear stress, $\bar{\tau}_{xz}$

Figure 13. Laminate H: through-thickness distribution of the normalised in-plane stress and transverse shear stress.

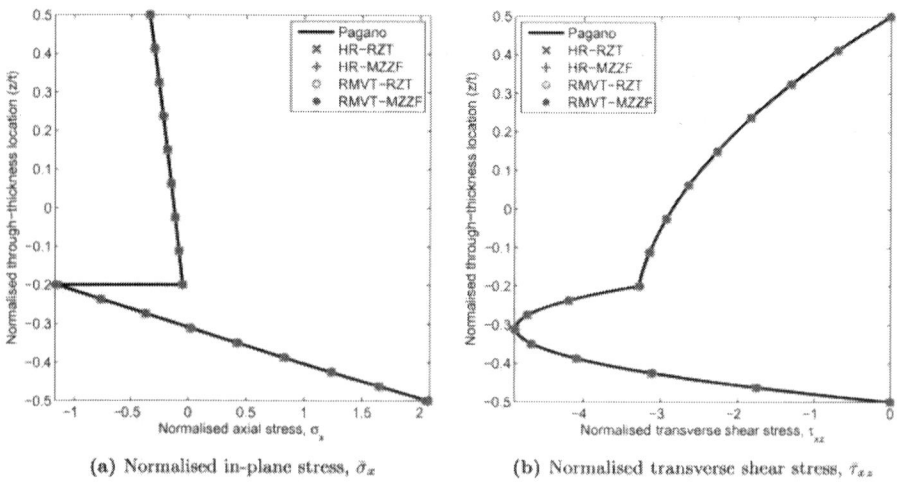

(a) Normalised in-plane stress, $\bar{\sigma}_x$

(b) Normalised transverse shear stress, $\bar{\tau}_{xz}$

Figure 14. Laminate I: through-thickness distribution of the normalised in-plane stress and transverse shear stress.

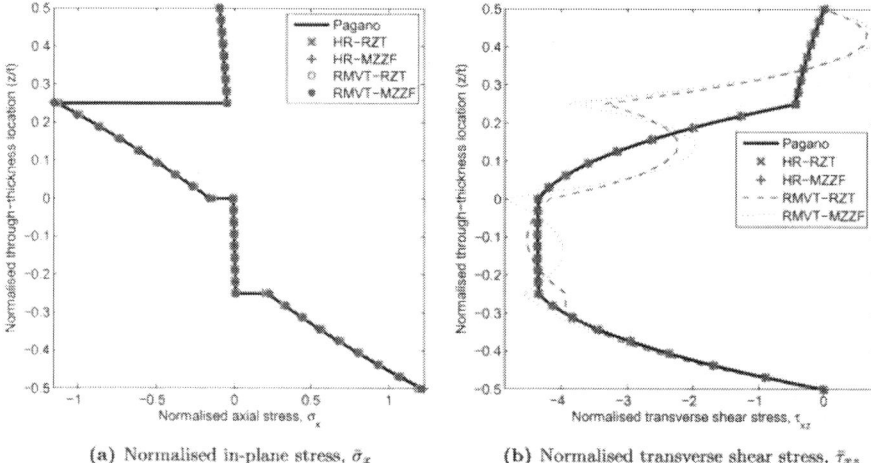

(a) Normalised in-plane stress, $\bar{\sigma}_x$

(b) Normalised transverse shear stress, $\bar{\tau}_{xz}$

Figure 15. Laminate J: through-thickness distribution of the normalised in-plane stress and transverse shear stress.

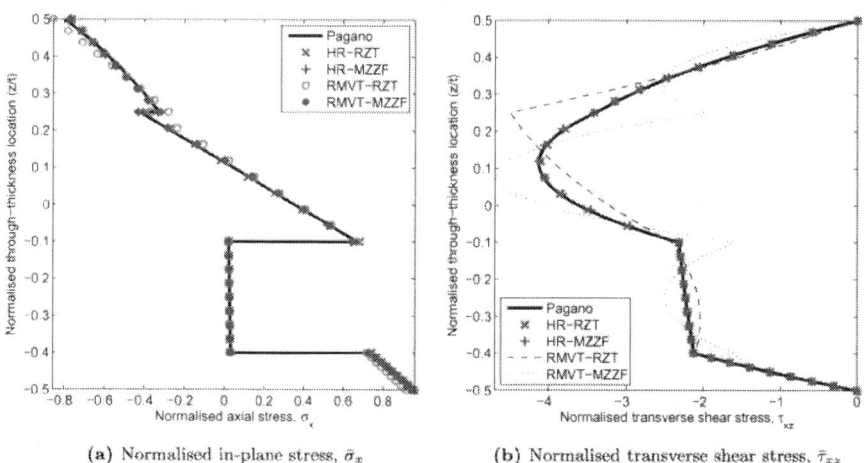

(a) Normalised in-plane stress, $\bar{\sigma}_x$

(b) Normalised transverse shear stress, $\bar{\tau}_{xz}$

Figure 16. Laminate K: through-thickness distribution of the normalised in-plane stress and transverse shear stress.

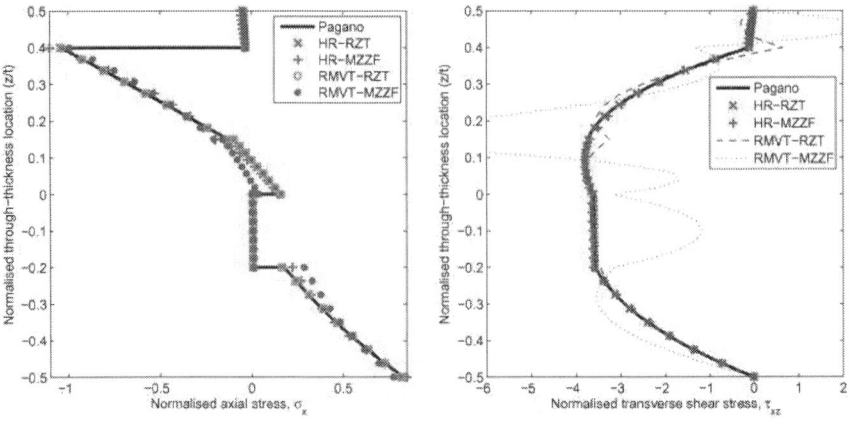

(a) Normalised in-plane stress, $\bar{\sigma}_x$ (b) Normalised transverse shear stress, $\bar{\tau}_{xz}$

Figure 17. Laminate L: through-thickness distribution of the normalised in-plane stress and transverse shear stress.

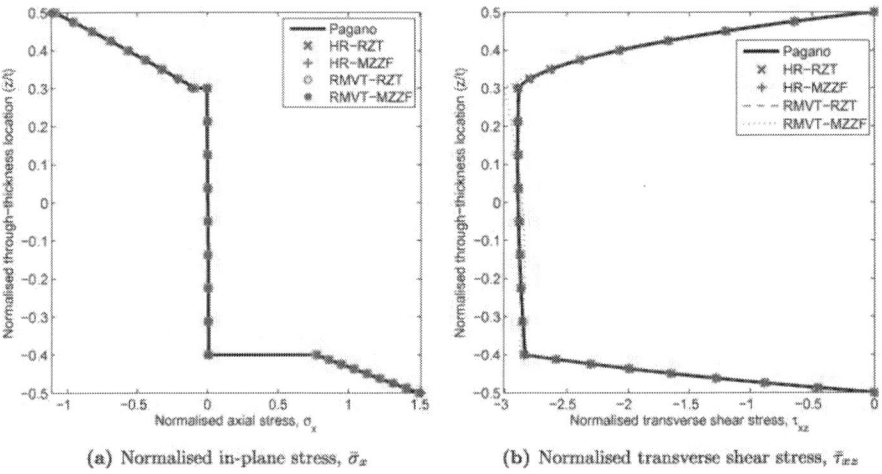

(a) Normalised in-plane stress, $\bar{\sigma}_x$ (b) Normalised transverse shear stress, $\bar{\tau}_{xz}$

Figure 18. Laminate M: through-thickness distribution of the normalised in-plane stress and transverse shear stress.

Table 3 shows that the accuracy of MRZZ for $\bar{w}$ and $\bar{\sigma}_x^{max}$ is within 1.1% for the zero B-matrix laminates considered here. For Laminates A-E the accuracy of the maximum transverse shear stress $\bar{\tau}_{xz}^{max}$ is also within 2.3%. Furthermore, the corresponding Fig. 6, Fig. 7, Fig. 8, Fig. 9 and Fig. 10 show that the axial stress and transverse shear stress for these laminates is accurately captured through the entire laminate thickness. The error in maximum transverse shear stress for the anti-symmetric Laminate H is slightly greater at 4.13%. However, the transverse shear profile in Fig.

13(b) shows that this error is on the conservative side and that the overall profile is captured well. Considering that MRZZ only requires two degrees of freedom, the spring modification factors $g^{(k)}$ and $e^{(k)}$ appear to be efficient methods of achieving both accurate axial and transverse shear stress solutions for zero B-matrix laminates. This result is especially noteworthy as most displacement-based models with the principle of virtual displacements do not obtain accurate transverse shear stresses *a priori* and require a post-processing stress recovery step (Carrera, 2003a).

However, the errors in $\bar{\tau}_{xz}^{max}$ for the honeycomb sandwich beams F and G are excessive. The corresponding Fig. 11 and Fig. 12(b) show that the MRZZ shear function is unable to capture the reversal of the transverse shear stress field in the stiffer face layers. This effect happens because the shear stress assumption Eq. (10) does not allow the magnitude of transverse shear stress to decrease between the outer surface and the laminate mid-plane. This behaviour only occurs for extreme cases of transverse orthotropy, when the transverse shear rigidity of an inner layer is insufficient to support the peak transverse shear stress of the adjacent outer layer. In essence, it is a load redistribution effect that occurs because the shear force must remain constant for any laminate under the same external load. Thus, as the transverse shear orthotropy between different layers is increased the transverse shear stress is increasingly shifted towards the stiffer layers. The extreme case of transverse orthotropy occurs when stiffer outer layers are essentially bending independently with fully reversed transverse shear profiles (i.e. the inner layer carries no transverse shear). To capture this effect an extra z-term with an appropriate coefficient would have to be added to the MRZZ shear stress assumption of Eq. (10). Fig. 11 and Fig. 12(b) also show that the shear force, being the integral through the thickness, is much less for the MRZZ formulation than for Pagano's solution. This occurs because the transverse shear stress profile is dominated by the smallest value of $G_{xz}^{(k)}$ in the laminate as a result of the reciprocal sum definition of G in Eq.(6).

On the other hand, the corresponding MRZZ axial stress fields for Laminates F and G inFig. 11 and Fig. 12(a) remain accurate throughout the entire thickness. Thus, more accurate transverse shear stress profiles for laminates F and G could be obtained in a post-processing stress recovery step although these would not satisfy the original beam equilibrium equations. Finally, the axial stress plots in Fig. 8, Fig. 12 and Fig. 13(a) show that the slope of the externally weak $90°$ layers is rigorously and accurately captured using MRZZ without the need for the RZT implementation rule described in Section 2.1.

For all analysed laminates the accuracy of HR-RZT is within 1% for all three metrics $\bar{W}$, $\bar{\sigma}_x^{max}$ and $\bar{\tau}_{xz}^{max}$. The corresponding through-thickness plots in Fig. 6, Fig. 7, Fig. 8, Fig. 9, Fig. 10, Fig. 11, Fig. 12, Fig. 13, Fig. 14, Fig. 15, Fig. 16, Fig. 17 and Fig. 18 show that both axial stress and transverse shear stress profiles are closely matched to Pagano's 3D elasticity solution for any type of laminate. Most importantly, the transverse shear stress profile is captured accurately from the *a priori* model assumption. Moreover, the through-thickness plots of $\bar{\sigma}_z$ in Fig. 19 and Fig. 20 show that the transverse normal stress field is also captured accurately. Thus, the HR-RZT formulation is shown to be an ESL theory with seven unknowns that provides 3D stress field predictions to within nominal errors of Pagano's 3D elasticity solution for arbitrarily laminated, thick (thickness-to-length ratio 1:8), anisotropic, composite and sandwich beams without the need for post-processing stress recovery steps.

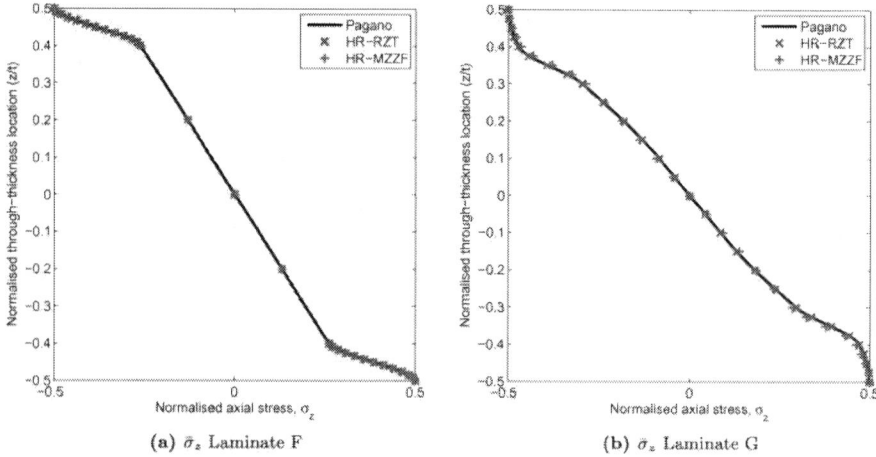

(a) $\bar{\sigma}_z$ Laminate F   (b) $\bar{\sigma}_z$ Laminate G

Figure 19. Through-thickness distribution of the normalised transverse normal stress for laminates F and G.

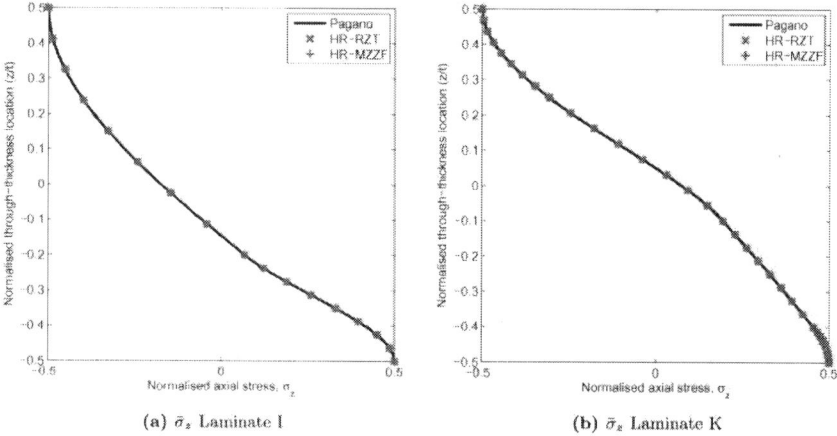

Figure 20. Through-thickness distribution of the normalised transverse normal stress for laminates I and K.

The accuracy of the HR-MZZF formulation is within the same range as HR-RZT for most laminates, and for cross-ply Laminates A,B,D and I the results are identical. Small discrepancies exist for cross-ply laminates Laminates C and J because of the presence of EWLs which are not taken account of in the MZZF. For laminates with at least three unique plies, arising either due to different material properties or varying ply orientations, the HR-MZZF formulation generally gives less accurate results for all three metrics metrics $\bar{W}$, $\bar{\sigma}_x^{max}$ and $\bar{\tau}_{xz}^{max}$ (Laminates E, F, G, H, K, L and M). For Laminates E, K and M the difference between the two theories is marginal, while for Laminates F, G, H and L the error in $\bar{\sigma}_x^{max}$ of HR-MZZF is an order of magnitude greater than for HR-RZT. In fact the HR-MZZF error in $\bar{\sigma}_x^{max}$ for Laminates F, G and L, and $\bar{W}$ for Laminate G is more than double the 3% threshold. Furthermore, Fig. 13 and Fig. 17(a) show that for Laminates H and L there are visible discrepancies in the $\bar{\sigma}_x$ through-thickness profile compared to Pagano's solution. The numerical errors in Table 3 and Table 4 suggest that HR-MZZF captures the maximum value of the transverse shear stress accurately for all laminates. However, the through-thickness shear stress profiles indicate that this is generally not the case. For example, in Fig. 12(b) the transverse shear stress is accurately captured in the outer $0°$ p-layers of Laminate G whereas there are visible discrepancies in the stress profile for all other layers. The authors want to emphasise that while the results of HR-MZZF are not as accurate as the HR-RZT formulation, overall the results are satisfactory given the challenging laminates considered here.

The HR *a priori* model assumption of transverse shear stress provides superior results to the model assumption in the RMVT formulation. The transverse shear stress profiles for laminates with a small number of layers, such as A, B, I and M, follow Pagano's solution closely for both RMVT-RZT and RMVT-MZZF. As the number of layers is increased the transverse shear profiles of both formulations oscillate around the 3D elasticity solution, most clearly shown in Fig. 9(b). In Laminates E, F, G, H and L the oscillations in the RMVT-MZZF solution significantly increases the maximum value of the transverse shear stress $\bar{\tau}_{xz}^{max}$ as indicated by the 2700% error for Laminate F in Table 3. In RMVT two independent assumptions are made for the displacement and transverse shear stress fields which are enforced to be kinematically compatible in the variational statement. However, as was shown in Section 2, the ZZ effect in the displacement field is directly related to the presence of C^1 discontinuous transverse shear strains that result from continuity requirements on transverse shear stresses at layer interfaces, and as such, the independence of the two fields is not absolute. Whereas the minimisation of the strain energy in RMVT guarantees that the geometric and assumed model shear strains are compatible there is no such condition on equilibrium between the axial stress and the transverse shear stress. In the HR principle the situation is reversed in that compatibility of geometric and assumed model strains is not guaranteed whereas equilibrium of stresses is enforced. In terms of deriving accurate stress fields, which are the critical measures for failure analyses, it seems that the enforcement of equilibrium is more critical than displacement compatibility, and hence HR seems to be a better formulation than RMVT.

For most laminates the through-thickness profiles of $\bar{\tau}_{xz}$ for RMVT-RZT show major discrepancies compared with Pagano's solution. On the other hand, the through-thickness plots of $\bar{\sigma}_x$ closely match Pagano's solution. Thus, the RMVT-RZT axial stress fields may be integrated in the equilibrium equations to compute more accurate transverse stresses, a step which was advised by Toledano & Murakami in their original papers on RMVT (Toledano and Murakami, 1987) and later reinforced by Carrera (2000). While the HR formulation introduced here features seven functional unknowns, the RMVT formulation only features six variables. Thus, the overall computational efficiency of the RMVT formulation with respect to the HR theory depends on the computational effort involved in the extra post-processing step.

It is also observed that for all EWL laminates the RMVT-RZT formulation considerably improves the accuracy of $\bar{\tau}_{xz}^{max}$ compared to RMVT-MZZF. As mentioned above, this occurs because the effects of EWLs are not taken into consideration within the MZZF. Furthermore, Fig. 11, Fig. 12 and Fig. 17(a) show significant discrepancies between the RMVT-MZZF through-thickness profiles of $\bar{\sigma}_x$ compared to Pagano's solution. Combined with the greater accuracy of HR-RZT compared to HR-MZZF, this corroborates the findings of Gherlone (2013) that the MZZF may lead to inferior results compared to the RZT ZZ functions for general laminates. In fact Toledano & Murakami point out that the "inclusion of the zig-zag shaped C^0 function was motivated by the displacement microstructure of *periodic* laminated composites" and that "for general laminate configurations, this periodicity is destroyed" such that the "theory should be expected to break down in these particular cases" (Toledano and Murakami, 1987). The present authors want to emphasise that the MZZF results in nominal errors for most practical laminates when employed in a third-order theory coupled with the Hellinger–Reissner mixed variational statement. However, for sandwich beams with very flexible cores or highly heterogeneous laminates the constitutive independence of the MZZF may lead to larger errors.

In conclusion, the HR-RZT formulation is the most accurate of the formulations investigated here for predicting bending deflections, axial bending stresses and transverse bending stresses from *a priori* model assumptions as a result of guaranteeing that stresses equilibrate. The RMVT-RZT theory provides similar accuracy for axial stresses but requires a *posteriori* stress recovery step for accurate transverse shear stresses. In terms of computational efficiency, there is a tradeoff between the extra degree of freedom in the HR-RZT formulation and the post-processing step in the RMVT-RZT formulation.

Assessment of Transverse Shear, Transverse Normal and Zig-zag Effects

Previous authors (Murakami, 1986, Tessler et al., 2009 and Carrera, 2004) have shown that ignoring the ZZ effect may lead to significant underestimations of the peak axial and transverse stresses. The inaccuracies are especially pronounced for sandwich beams because of the large degree of transverse orthotropy between the flexible core and stiff face layers. Even though a relatively large thickness-length ratio of 1:8 was analysed in this study the ZZ effect is important for longer beams as well. Furthermore, a major aim of sandwich construction is to separate the stiff face layers as far as possible for maximum bending stiffness. This means that sandwich panels are often of larger thickness-to-length ratios than most other laminated structures.

However, the ZZ effects for practical composite laminates may not be as significant. Many industrial lamination guidelines prevent the use of thick blocks of same orientation plies to prevent problems associated with transverse cracking. As such Laminates A–C are not representative of typical laminates used in practice and the results for Laminate D in Fig. 9 show that dispersing the $0°$ and $90°$ layers through the thickness greatly reduces the ZZ effect.

The relative effects of transverse shear deformation, transverse normal deformation and the ZZ effect may be analysed numerically using the bending displacement magnitude of the Hellinger–Reissner theory. For simplicity, we only examine symmetric laminates ($N=O=0$), and ignore the effect of the higher-order moment P . Also the entire surface traction acts on the top surface such that $\hat{P}_b = 0$ and $\hat{P}_t = q_0 \sin \pi x/a$. To start, the effect of the ZZ deformation is also ignored such that the relative importance of transverse shear and transverse normal deformation can be compared. In this case, the governing Eqs. (50a), (50b) and (50c) of the HR theory reduce to

$$M_{,xx} + \hat{P}_t = 0 \tag{59a}$$

$$s^{CLA}M - (\bar{\eta} - \bar{\omega})M_{,xx} + \bar{\rho}M_{,xxxx} + \lambda_{2,xx} = 0 \tag{59b}$$

and boundary condition Eq. (51b)

$$\frac{\bar{\omega}}{2}M + \bar{\rho}M_{,xx} + \lambda_2 = w \tag{60}$$

is required to calculate the bending magnitude w_0 from M and λ_2. Using the solution assumption in Eq. (56a) results in

$$w_0 = \frac{a^4}{\pi^4} \left[s^{CLA} + \frac{\pi^2}{a^2} \left(\bar{\eta} - \frac{\bar{\omega}}{2} \right) \right] q_0 \tag{61}$$

where $sCLA = 1/DCLA$,   and $\bar{\eta}$, $\bar{\omega}$ and $\bar{\rho}$ are   equal to η_{22}, ω_{22} and ρ_{22} respectively, when calculated without the influence of terms related to N,O,P and L . Note, that the normal shear correction factor $\bar{\rho}$ does not influence the bending deflection w_0 result in Eq. (61). The bending deflection can be non-dimensionalised into three separate entities by dividing by the factor $q_0 sCLA a^4/\pi^4$,

$$\bar{w}_0 = \bar{w}_{CLA} + \bar{w}_{TS} + \bar{w}_{TN} \quad \text{where}$$

$$\bar{w}_{CLA} = 1, \quad \bar{w}_{TS} = \frac{\bar{\eta}}{s^{CLA}} \frac{\pi^2}{a^2}, \quad \bar{w}_{TN} = -\frac{\bar{\omega}}{2s^{CLA}} \frac{\pi^2}{a^2} \tag{62}$$

where $\bar{w}_{TS}$ and $\bar{w}_{TN}$ refer to transverse shear and transverse normal deflection, respectively. These three factors are plotted against the thickness to length ratio (t/a) in Fig. 21 for Laminate D. Furthermore, this plot shows a comparison between the total normalised deflection $\bar{w}_0$ and Pagano's normalised solution $\bar{w}_{pag}$. Laminate D is chosen here to minimise the ZZ effect on Pagano's 3D elasticity solution and allow a fair comparison with $\bar{w}_0$.

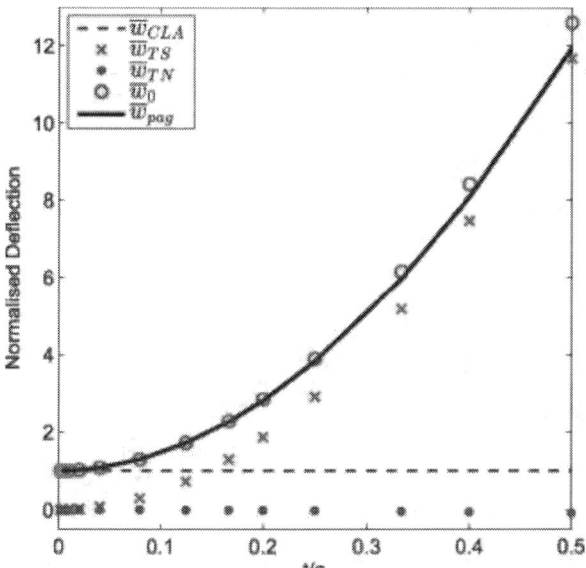

Figure 21. Laminate D: change in non-dimensional CLA, transverse shear and transverse normal bending deflection components versus thickness to length ratio t/a.

Fig. 21 also shows the parabolic relationship of $\bar{w}_{TS}$ with respect to the beam thickness ratio t/a. Furthermore, the transverse shear component plays a more significant role than the transverse normal deflection. It can be observed that the transverse normal component is actually negative, i.e. it stiffens the structure. This arises because $\bar{w}_{TN}$ only captures the Poisson's effect of σ_z on the bending deformation as indicated by the presence of the Poisson's effect correction factor $\bar{\omega}$. If normal compressibility effects are to be included their effects have to be factored into the initial assumption of the displacement field in

Eq. (19a) and (19b) that underlies the derivation of the theory. Finally, the comparison between $\overline{w}_0$ and $\overline{w}_{pag}$ shows that the HR model maintains accurate predictions of the bending deflection up to very deep aspect ratios of $t/a=0.4$.

Fig. 21 shows that the effect of $\overline{w}_{TN}$ is insignificant compared to the magnitude of $\overline{w}_{TS}$ for a $[0/(90/0)_{25}]$ laminate. Although the transverse normal correction factor $\bar{\omega}$ varies with layup, a numerical study showed that for practical carbon- and glass-reinforced composite and sandwich panels the magnitude of this factor is always negligible compared to the transverse shear factor $\bar{\eta}$.

Therefore, the transverse normal component is ignored in the assessment of the influence of the ZZ effect on the bending behaviour. From the boundary condition Eq.(51b) it is seen that Lagrange multiplier $\lambda_2 = w$ when all transverse normal correction factors vanish. As such, when ZZ effects are included the bending deflection equals

$$w_0^{ZZ} = w_{CLA}^{ZZ} + w_{TS}^{ZZ} \quad \text{where}$$

$$w_{CLA}^{ZZ} = \left[\frac{a^4}{\pi^4}\bar{s}_{22} - \frac{\left(\bar{s}_{25}\frac{a^2}{\pi^2} + \bar{\eta}_{25}\right)^2}{\bar{s}_{55} + \bar{\eta}_{55}\frac{\pi^2}{a^2}}\right]q_0 \tag{63a}$$

$$w_{TS}^{ZZ} = \bar{\eta}_{22}\frac{a^2}{\pi^2}q_0 \tag{63b}$$

where $\bar{s}_{ij}$ and $\bar{\eta}_{ij}$ are equal to s_{ij} and η_{ij} respectively, when calculated without the influence of terms related to $N, O,$ and P. The bending deflection is non-dimensionalised using the factor $q_0 s^{CLA}a^4/\pi^4$, thus

$$\bar{w}_{CLA}^{ZZ} = \frac{\bar{s}_{22}}{s^{CLA}} - \frac{\left(\bar{s}_{25} + \bar{\eta}_{25}\frac{\pi^2}{a^2}\right)^2}{s^{CLA}\left(\bar{s}_{55} + \bar{\eta}_{55}\frac{\pi^2}{a^2}\right)} \tag{64a}$$

$$\bar{w}_{TS}^{ZZ} = \frac{\pi^2}{a^2}\frac{\bar{\eta}_{22}}{s^{CLA}}. \tag{64b}$$

Note that the bending flexibility s^{CLA} of CLA and s_{22} of RZT are generally not equal. The terms are related by a constant of proportionality that is independent of the laminate thickness t. The two components $\bar{w}_{CLA}^{ZZ}$ and $\bar{w}_{TS}^{ZZ}$ in Eq. (64a) and (64b) can be compared to the corresponding bending components that ignore the ZZ effect in Eq. (62).

Thus the total deflection ratio $r_w = \overline{W}_0^{ZZ} / \overline{W}_0$ and shear deflection ratio $r_{TS} = \overline{W}_{TS}^{ZZ} / \overline{W}_{TS}$ are used as metrics to assess the influence of the ZZ effect on the bending displacement.

Fig. 22(a) shows that the ratio of transverse shear components is invariant with t/a but may vary with the stacking sequence. Furthermore, the ZZ effect always reduces the magnitude of transverse shear deformation. Fig. 22(b) shows that the ZZ effect reduces the overall bending deflection of all analysed laminates and that this effect is greatest for the two honeycomb core sandwich beams F and G, i.e. stiffening is most for laminates with the greatest ZZ effect. Furthermore, the reduction in bending displacement is highly non-linear in t/a and converges to a constant value as the thickness of the beam approaches the length. The authors believe that this stiffening effect occurs partially due to the lower sensitivity to transverse shear deformation (Fig. 22(a)) and due to the action of the higher-order moment L . In general, the sum of the two non-dimensionalised coefficients $\overline{W}_{CLA}^{ZZ}$ and $\overline{W}_{TS}^{ZZ}$ may be used to gauge the ZZ effect on a specific combined laminate and thickness to length ratio case.

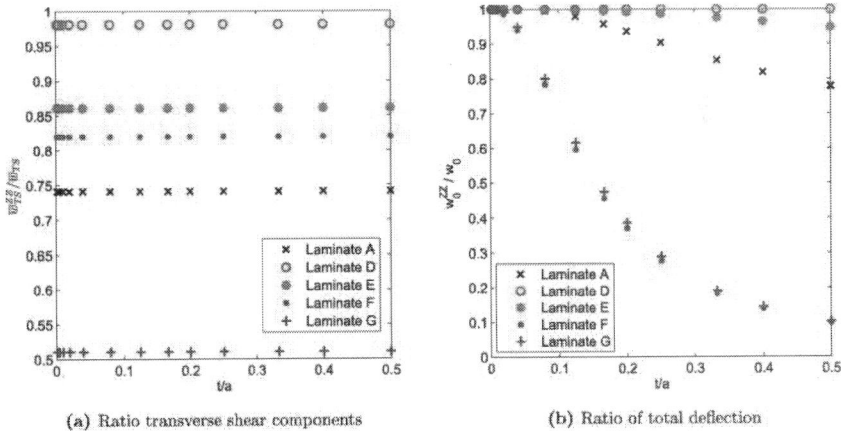

(a) Ratio transverse shear components   (b) Ratio of total deflection

Figure 22. Ratio of transverse shear components and total deflection, as calculated from HR-RZT theory with and without ZZ effects, versus thickness to length ratio (t/a) for different laminates in Table 2.

CONCLUSIONS

In this paper, the fundamental mechanics driving the zig-zag (ZZ) effect in multilayered structures was analysed. The ZZ effect was attributed to differences in transverse shear strains at layer interfaces that require changes in slope of the deformation field in order to satisfy the kinematic equations. The dual requirement of transverse shear stress and displacement continuity at layer interfaces led to the notion of modelling the transverse shear mechanics of a multi-layered structure using a combination of "springs-in-series" and "springs-in-parallel" systems. By defining equivalent spring stiffnesses for the transverse shear modulii and axial modulii, Reddy's parabolic shear function was modified using a layerwise curvature modification factor. This modifed Reddy zig-zag theory (MRZZ), guarantees that the shear stress assumption satisfies the layer interface continuity conditions, traction free surface conditions and that laminates with Externally Weak Layers (EWL) can be modelled in an appropriate manner. However, the MRZZ theory is only applicable to laminates with zero B-matrix terms. Furthermore, the notion that accurate transverse stresses can be obtained by integrating the axial stress in Cauchy's equilibrium equations led to the development of a second theory using the Hellinger–Reissner (HR) variational principle. In this theory the ZZ function of the refined zig-zag theory (RZT) introduced by Tessler et al. (2007) and Murakami's ZZ function (MZZF) were used to develop two alternative theories HR-RZT and HR-MZZF, respectively. The governing equations for the MRZZ and HR formulations were derived for laminated beams under the plane-strain condition and the theories were validated with respect to Pagano's 3D elasticity solution.

The results for different laminated composite and sandwich beams with zero B-matrix terms show that the bending deflection and axial stress field is captured to within 1.1% percent by the MRZZ theory. MRZZ is also capable of making accurate predictions of the transverse shear stress field from constitutive equations when the profile does not cause local peaks at z-wise positions other than the laminate mid-plane. This latter effect occurs for extreme cases of transverse orthotropy, i.e. when the transverse shear rigidity of an inner layer is insufficient to support the peak transverse shear stress of the adjacent outer layer. In these cases an additional *posteriori* stress recovery step should be performed for accurate transverse stress prediction. Nevertheless, for practical, symmetrically laminated composite laminates MRZZ provides accurate transverse shear stress fields that satisfy interfacial continuity conditions directly from the constitutive equations at the same computational cost as the Timoshenko theory.

The HR-RZT is the best performing theory considered, predicting the bending deflection and three stress fields σ_x, τ_{xz} and σ_z to within 1% of Pagano's solution even for highly heterogeneous laminates with arbitrary thickness ratios, ply material orientations and layer material properties. This result is noteworthy because the transverse shear and transverse normal stresses are captured accurately without requiring a post-processing stress recovery step. In the HR principle the equilibrium of in-plane and transverse stress fields is enforced in a weak sense such that boundary layers near clamped and free edges, and accurate transverse stresses can be derived directly from the plate equations. Furthermore, by using the same variables for in-plane and transverse stress fields the computational effort is kept low.

The results of the HR-MZZF theory showed that even though the MZZF is not based on the constitutive heterogeneity of a laminate it captures the three-dimensional stress field to similar accuracy of HR-RZT. For some laminates the errors in HR-MZZF compared to Pagano's solution are as great as 10% whereas the HR-RZT formulation remains to within 1% (Laminate G). The performance of the HR formulations was also compared to corresponding theories developed using the Reissner Mixed Variational Theory (RMVT). Whereas the RMVT-RZT and RMVT-MZZF give accurate predictions for the bending deflection and axial stress, the model assumptions for transverse shear stress may be highly inaccurate when the number of layers exceeds three. As a result, the RMVT formulations require extra post-processing steps to guarantee accurate transverse stress results. However, compared to the HR formulation the RMVT formulation reduces the variable count by one. Thus, the overall computational efficiency of the RMVT formulation with respect to the HR theory depends on the effort involved in the extra post-processing step.

Finally, for practical non-sandwich beams used in industry the ZZ effect does not seem to be of great importance as many lamination codes prohibit thick blocks of $0°$ and $90°$ plies. In these cases higher-order effects such as "stress-channelling" are more important. Furthermore, two non-dimensional factors have been identified that allow the influence of the ZZ effect on the classical bending deflection and transverse shear behaviour to be quantified. The results show that including the ZZ effect in the model reduces the effect of transverse shear deformation and generally acts to stiffen the structure.

ACKNOWLEDGEMENTS

This work was supported by the Engineering and Physical Sciences Research Council through the EPSRC Centre for Doctoral Training in Advanced Composites for Innovation and Science [Grant No. EP/G036772/1].

Appendix A. Transverse shear stress and displacement continuity constants

The assumed transverse shear profile for MRZZ is given by,

$$\tau_{xz}^{(k)} = G\left\{A^{(k)} - \frac{4}{t^2}m^{(k)}z^2\right\}\bar{\gamma}_{xz}.$$

The transverse shear stress must vanish at the bottom interface. Thus,

$$\tau_{xz}^{(1)}(z_0) = 0 \Rightarrow A^{(1)} = \frac{4}{t^2}m^{(1)}z_0^2$$

where superscripts pertain to layerwise quantities and subscripts to interfacial quantities as shown in Fig. 1. Similarly the interface continuity conditions have to be satisfied,

$$\tau_{xz}^{(k)}(z_k) = \tau_{xz}^{(k+1)}(z_k), \quad k = 1\ldots N-1$$

$$A^{(k)} - A^{(k+1)} = \frac{4}{t^2}(m^{(k)} - m^{(k+1)})z_k^2, \quad k = 1\ldots N-1.$$

This results in N algebraic equations which can be solved simultaneously to find the N layerwise constants $A^{(k)}$, $k = 1\ldots N$.

The layerwise constants $c^{(k)}$ are found by enforcing continuity of displacements at layer interfaces and the condition that $u_x(x,0)=u_0$. This condition gives two possible solutions for the layerwise constant $c^{(k)}$. If the reference surface $z=0$ is located within a layer k_0 the constants are given by

$$c^{(k_0)}=0 \tag{A.2a}$$

$$c^{(k)} = \sum_{i=k_0+1}^{k} \left[\left(g^{(i-1)} A^{(i-1)} - g^{(i)} A^{(i)} \right) z_{i-1} - \frac{4}{3t^2} \left(g^{(i-1)} m^{(i-1)} - g^{(i)} m^{(i)} \right) z_{i-1}^3 \right], \quad k > k_0$$

(A.2b)

$$c^{(k)} = \sum_{i=k}^{k_0-1} \left[\left(g^{(i+1)} A^{(i+1)} - g^{(i)} A^{(i)} \right) z_{i} - \frac{4}{3t^2} \left(g^{(i+1)} m^{(i+1)} - g^{(i)} m^{(i)} \right) z_{i}^3 \right], \quad k < k_0.$$

(A.2c)

Similarly, if the reference surface $z=0$ lies on the interface between two layers, defined as k^+ and k^-, the constants are given by

$$c^{(k+)} = c^{(k-)} = 0$$

(A.3a)

$$c^{(k)} = \sum_{i=k^++1}^{k} \left[\left(g^{(i-1)} A^{(i-1)} - g^{(i)} A^{(i)} \right) z_{i-1} - \frac{4}{3t^2} \left(g^{(i-1)} m^{(i-1)} - g^{(i)} m^{(i)} \right) z_{i-1}^3 \right], \quad k > k^+$$

(A.3b)

$$c^{(k)} = \sum_{i=k}^{k^--1} \left[\left(g^{(i+1)} A^{(i+1)} - g^{(i)} A^{(i)} \right) z_{i} - \frac{4}{3t^2} \left(g^{(i+1)} m^{(i+1)} - g^{(i)} m^{(i)} \right) z_{i}^3 \right], \quad k < k^-.$$

(A.3c)

Appendix B. Derivation of HR governing equations

The Hellinger–Reissner function in Eq. (47) can be split into separate components representing the potential of axial stress $\Pi_{\sigma x}$, transverse shear stress $\Pi_{\tau} xz$ and transverse normal $\Pi_{\sigma z}$ stress, the potential of boundary tractions Π_{Γ} and the potential of the Lagrange multiplier constraints Π_{λ}. Substituting the pertinent expressions for stresses and strain into the functional of Eq. (47) yields,

$$\delta\Pi = \delta(\Pi_{\sigma x} + \Pi_{\tau} xz + \Pi_{\sigma z} + \Pi_{\lambda} + \Pi_{\Gamma}) = 0$$

$$\Pi_{\sigma x} = \frac{1}{2} \int_V \sigma_x^T \epsilon_x dV = \frac{1}{2} \int_V \mathcal{F}^T s^T f_{\phi}^{(k)T} \bar{Q}^{(k)} f_{\phi}^{(k)} s \mathcal{F} dV$$

(B.1a)

$$\Pi_{\tau_{xz}} = \frac{1}{2} \int_V \tau_{xz}^T \gamma_{xz} dV$$
$$= \frac{1}{2} \int_V \left(\boldsymbol{c}^{(k)} \boldsymbol{s} \mathcal{F}_x + \hat{T}_b \right)^T \frac{1}{G_{xz}^{(k)}} \left(\boldsymbol{c}^{(k)} \boldsymbol{s} \mathcal{F}_x + \hat{T}_b \right) dV$$

(B.1b)

$$\Pi_{\sigma_z} = \frac{1}{2} \int_V \sigma_z^T \epsilon_z dV$$
$$= \frac{1}{2} \int_V \left(\boldsymbol{e}^{(k)} \boldsymbol{s} \mathcal{F}_{xx} - \hat{T}_{b,x}(z - z_0) + \hat{P}_b \right)^T$$
$$\cdot \left[R_{13}^{(k)} \bar{Q}^{(k)} \boldsymbol{f}_\phi^{(k)} \boldsymbol{s} \mathcal{F} + R_{33}^{(k)} \left(\boldsymbol{e}^{(k)} \boldsymbol{s} \mathcal{F}_{xx} - \hat{T}_{b,x}(z - z_0) + \hat{P}_b \right) \right] dV$$

(B.1c)

$$\Pi_\lambda = \int \lambda_1 \left(N_x + \hat{T}_t - \hat{T}_b \right) dx$$
$$+ \int \lambda_2 \left(M_{xx} + z_N \hat{T}_{tx} - z_0 \hat{T}_{b,x} + \hat{P}_t - \hat{P}_b \right) dx$$

(B.1d)

$$\Pi_\Gamma = - \int_{S_2} \left(\sigma_x \hat{u}_x^{(k)} + \tau_{xz} \hat{w} \right) dS_2 = - \int \left[\sigma_x \boldsymbol{f}_\phi^{(k)} \hat{u} + \tau_{xz} \hat{w} \right] \Big|_{x_A}^{x_B} dz$$

(B.1e)

Performing the variations on the functionals in Eqs. (B.1a), (B.1b), (B.1c),(B.1d) and (B.1e) following the rules of the calculus of variations results in the following expressions. For the potential of axial stress we have,

$$\delta\Pi_{\sigma_x} = \delta\left\{ \frac{1}{2} \int \mathcal{F}^T \boldsymbol{s}^T \left(\int \boldsymbol{f}_\phi^{(k)^T} Q^{(k)} \boldsymbol{f}_\phi^{(k)} dz \right) \boldsymbol{s} \mathcal{F} dx \right\}$$
$$= \delta\left\{ \frac{1}{2} \int \mathcal{F}^T \boldsymbol{s}^T S^T \boldsymbol{s} \mathcal{F} dx \right\} = \delta\left\{ \frac{1}{2} \int \mathcal{F}^T \boldsymbol{s} \mathcal{F} dx \right\} = \int \mathcal{F}^T \boldsymbol{s} \delta \mathcal{F} dx$$

(B.2)

For the potential of transverse shear stress,

$$\delta\Pi_{\tau_{xz}} = \delta\left\{\frac{1}{2}\int\left[\mathcal{F}_x^T\mathbf{s}^I\left(\int \mathbf{c}^{(k)^I}\frac{1}{G_{xz}^{(k)}}\mathbf{c}^{(k)}dz\right)\mathbf{s}\mathcal{F}_x\right.\right.$$
$$\left.\left.+2\hat{T}_b\left(\int\frac{1}{G_{xz}^{(k)}}\mathbf{c}^{(k)}dz\right)\mathbf{s}\mathcal{F}_x + \int\frac{\hat{T}_b^2}{G_{xz}^{(k)}}dz\right]dx\right\} = \int\left(\mathcal{F}_x^T\boldsymbol{\eta}+\hat{T}_b\boldsymbol{\chi}\right)\delta\mathcal{F}_x dx$$

(B.3)

where $\boldsymbol{\eta}$ is a 5×5 matrix of shear coefficients that automatically includes pertinent shear correction factors, and $\boldsymbol{\chi}$ is a 1×5 row vector of correction factors that enforce transverse shearing effects of the surface shear tractions. Matrix $\boldsymbol{\eta}$ and row vector $\boldsymbol{\chi}$ are defined by

$$\boldsymbol{\eta} = \mathbf{s}^T\left(\int \mathbf{c}^{(k)^I}\frac{1}{G_{xz}^{(k)}}\mathbf{c}^{(k)}dz\right)\mathbf{s}$$

(B.4a)

$$\boldsymbol{\chi} = \left(\int\frac{1}{G_{xz}^{(k)}}\mathbf{c}^{(k)}dz\right)\mathbf{s}$$

(B.4b)

Performing integration by parts on Eq. (B.3) results in,

$$\delta\Pi_{\tau_{xz}} = \left[\mathcal{F}_x^T\boldsymbol{\eta}+\hat{T}_b\boldsymbol{\chi}\right]\Big|_{x_A}^{x_B} - \int\left(\mathcal{F}_{xx}^T\boldsymbol{\eta}+\hat{T}_{b,x}\boldsymbol{\chi}\right)\delta\mathcal{F}dx$$

(B.5)

For the potential of transverse normal stress we expand the parantheses in Eq. (B.1c), define the following transverse normal correction factors

$$\boldsymbol{\omega} = \mathbf{s}^T\left(\int \mathbf{e}^{(k)^I}R_{13}^{(k)}\bar{Q}^{(k)}\boldsymbol{f}_\phi^{(k)}dz\right)\mathbf{s}$$

(B.6a)

$$\boldsymbol{\omega}_t = \left(\int R_{13}^{(k)}(z-z_0)\bar{Q}^{(k)}\boldsymbol{f}_\phi^{(k)}dz\right)\mathbf{s}$$

(B.6b)

$$\boldsymbol{\omega}_p = \left(\int R_{13}^{(k)}\bar{Q}^{(k)}\boldsymbol{f}_\phi^{(k)}dz\right)\mathbf{s}$$

(B.6c)

$$\rho = \mathbf{s}^T\left(\int \mathbf{e}^{(k)^I}R_{33}^{(k)}\mathbf{e}^{(k)}dz\right)\mathbf{s}$$

(B.6d)

$$\boldsymbol{\rho}_t = \left(\int R_{33}^{(k)}(z - z_0)e^{(k)}dz \right)s \tag{B.6e}$$

$$\boldsymbol{\rho}_p = \left(\int R_{33}^{(k)}e^{(k)}dz \right)s \tag{B.6f}$$

where ω and ρ are 5×5 matrixes and $\omega_t,\ \omega_p,\ \rho_t$ and ρ_p are 1×5 row vectors, and integrate by parts to give

$$
\begin{aligned}
\delta\Pi_{\sigma_z} = \int &\left[\mathcal{F}_{,xxx}^T\rho + \mathcal{F}_{,xx}^T\omega - \hat{T}_{b,xxx}\rho_t + \hat{P}_{b,xx}\rho_p - \hat{T}_{b,x}\frac{\omega_t}{2} + \hat{P}_b\frac{\omega_p}{2} \right]\delta\mathcal{F}dx \\
&+ \left[\mathcal{F}_{,xx}^T\rho + \frac{\omega}{2}\mathcal{F}^T - \hat{T}_{b,x}\rho_t + \hat{P}_b\rho_p \right]\delta\mathcal{F}_{,x}\Big|_{x_A}^{x_B} \\
&- \left[\mathcal{F}_{,xxx}^T\rho + \frac{\omega}{2}\mathcal{F}_{,x}^T - \hat{T}_{b,xx}\rho_t + \hat{P}_{b,x}\rho_p \right]\delta\mathcal{F}\Big|_{x_A}^{x_B}
\end{aligned}
\tag{B.7}
$$

Finally, the potential of the Lagrange multipliers and the potential of contour loads are given by,

$$
\begin{aligned}
\delta\Pi_\lambda = \int &\left(N_{,x} + \hat{T}_t - \hat{T}_b \right)\delta\lambda_1 dx - \int \lambda_{1,x}\delta N dx + \lambda_1\delta N\Big|_{x_A}^{x_B} \\
&+ \int \left(M_{,xx} + z_N\hat{T}_{t,x} - z_0\hat{T}_{b,x} + \hat{P}_t - \hat{P}_b \right)\delta\lambda_2 dx + \int \lambda_{2,xx}\delta M \\
&+ \lambda_2\delta M_{,x}\Big|_{x_A}^{x_B} - \lambda_{2,x}\delta M\Big|_{x_A}^{x_B}
\end{aligned}
\tag{B.8}
$$

$$\delta\Pi_\Gamma = -\left[\delta\mathcal{F}\cdot\hat{\mathcal{U}} + \delta Q\hat{w} \right]\Big|_{x_A}^{x_B} = -\left[\delta\mathcal{F}\cdot\hat{\mathcal{U}} + \delta M_{,x}\hat{w} \right]\Big|_{x_A}^{x_B} \qquad q \tag{B.9}$$

The integral expressions in Eqs. (B.2), (B.5), (B.7), (B.8) and (B.9) combine to form the governing field Eqs. (50a), (50b) and (50c), while the terms evaluated at $x=x_A$ and $x=x_B$ combine to form the governing boundary Eq. (51a) and (51b). These equations feature three column vectors $\Lambda eq, \Lambda bc_1, \Lambda bc_2$ that include the Lagrange multipliers λ_1, λ_2 and their derivatives. These are given by,

$$\Lambda_{eq} = \begin{pmatrix} -\lambda_{1,x} \\ \lambda_{2,xx} \\ 0 \\ 0 \\ 0 \end{pmatrix}, \quad \Lambda_{bc1} = \begin{pmatrix} \lambda_1 \\ -\lambda_{2,x} \\ 0 \\ 0 \\ 0 \end{pmatrix}, \quad \Lambda_{bc2} = \begin{pmatrix} 0 \\ \lambda_2 \\ 0 \\ 0 \\ 0 \end{pmatrix} \tag{B.10}$$

Finally, the boundary displacement $\hat{W}$ in Eq. (B.9) is contained in the vector $\mathcal{W} = \begin{bmatrix} 0 & \hat{w} & 0 & 0 & 0 \end{bmatrix}^T$.

REFERENCES

Ambartsumyan, S.A., 1958a. On theory of bending plates. Isz. Otd. Tech. Nauk. AN SSSR 5, 69–77.

Ambartsumyan, S.A., 1958b. On a general theory of anisotropic shells. Prikl. Mat. Mekh. 22, 226–237.

Averill, R.C., 1994. Static and dynamic response of moderately thick laminated beams with damage. Compos. Eng. 4 (4), 381–395.

Averill, R.C., Yip, Y.C., 1996. Development of simple, robust finite elements based on refined theories for thick laminated beams. Compos. Struct. 59 (3), 3529–3546.

Barut, A., Madenci, E., Heinrich, J., Tessler, A., 2001. Analysis of thick sandwich construction by a 3,2-order theory. Int. J. Solids Struct. 38 (34), 6063–6067.

Barut, A., Madenci, E., Tessler, A., 2012. A refined zigzag theory for laminated composite and sandwich plates incorporating thickness stretch deformation. In: Proceedings of the 53rd AIAA/ASME/ASCE/AHS/ASC Structures, Structural Dynamics, and Materials Conference, Honolulu, Hawaii, USA.

Batra, R.C., Vidoli, S., 2002. Higher-order piezoelectric plate theory derived from a three-dimensional variational principle. AIAA J. 40 (1), 91–104.

Batra, R.C., Vidoli, S., Vestroni, F., 2002. Plane wave solutions and modal analysis in higher order shear and normal deformable plate theories. J. Sound Vib. 257 (1), 63–88.

Brischetto, S., Carrera, E., Demasi, L., 2009a. Free vibration of sandwich plates and shells by using zig-zag function. Shock Vib. 16, 495–503.

Brischetto, S., Carrera, E., Demasi, L., 2009b. Improved bending analysis of sandwich plates using zig-zag functions. Compos. Struct. 89, 408–415.

Carrera, E., 2000. A priori vs. a posteriori evaluation of transverse stresses in multilayered orthotropic plates. Compos. Struct. 48 (4), 245–260.

Carrera, E., 2001. Developments, ideas and evaluations based upon Reissner's mixed variational theorem in the modeling of multilayered plates and shells. ASME Appl. Mech. Rev. 54 (4), 301–329.

Carrera, E., 2002. Theories and finite elements for multilayered, anisotropic, composite plates and shells. Arch. Comput. Methods Eng. 9 (2), 87–140.

Carrera, E., 2003a. Historical review of zig-zag theories for multilayered plates and shells. Appl. Mech. Rev. 56 (3), 287–308.

Carrera, E., 2003b. Theories and finite elements for multi-layered plates and shells: a unified compact formulation with numerical assessment and benchmarking. Arch. Comput. Methods Eng. 10 (3), 5216–5296.

Carrera, E., 2004. On the use of the Murakami's zig-zag function in the modeling of layered plates and shells. Comput. Struct. 82 (7–8), 541–554.

Carrera, E., Filippi, M., Zappino, E., 2013a. Free vibration analysis of laminated beam by polynomial, trigonometric, exponential and zig-zag theories. J. Compos. Mater., 1–18

Carrera, E., Filippi, M., Zappino, E., 2013b. Laminated beam analysis by polynomial, trigonometric, exponential and zig-zag theories. Eur. J. Mech./A Solids 41, 58– 69.

Cook, G., Tessler, A., 1998. A 3,2-order bending theory for laminated composite and sandwich beams. Compos. Part B 29B, 565–576.

Cosentino, E., Weaver, P.M., 2010. An enhanced single-layer variational formulation for the effect of transverse shear on laminated orthotropic plates. Eur. J. Mech. A/Solids 29, 567–590.

Demasi, L., 2008. 13 hierarchy plate theories for thick and thin composite plates: the generalized unified formulation. Compos. Struct. 84, 256–270.

Demasi, L., 2012. Partially zig-zag advanced higher order shear deformation theories based on the generalized unified formulation. Compos. Struct. 94 (2), 363–375.

Di Sciuva, M., 1984. A refinement of the transverse shear deformation theory for multilayered orthotropic plates. L'aerotecnica Missile Spazio 62, 84–92.

Di Sciuva, M., 1985. Development of an anisotropic, multilayered, shear-deformable rectangular plate element. Compos. Struct. 21 (4), 789–796.

Everstine, G.C., Pipkin, A.C., 1971. Stress channelling in transversely isotropic elastic composites. Z. Angew. Math. Phys. (ZAMP) 22, 825–834.

Ferreira, A.J.M., Roque, C.M.C., Jorge, R.M.N., 2005. Analysis of composite plates by trigonometric shear deformation theory and multiquadrics. Comput. Struct. 83, 2225–2237.

Gherlone, M., 2013. On the use of zigzag functions in equivalent single layer theories for laminated composite and sandwich beams: a comparative study and some observations on external weak layers. J. Appl. Mech. 80, 1–19.

Groh, R.M.J., Weaver, P.M., 2015. Static inconsistencies in certain axiomatic higherorder shear deformation theories for beams, plates and shells. Compos. Struct. 120, 231–245.

Groh, R.M.J., Weaver, P.M., White, S., Raju, G., Wu, Z., 2013. A 2D equivalent singlelayer formulation for the effect of transverse shear on laminated plates with curvilinear fibres. Compos. Struct. 100, 464–478.

Jones, R.M., 1998. Mechanics of Composite Materials, second ed. Taylor & Francis Ltd., London, UK.

Karama, M., Abou Harb, B., Mistou, S., Caperaa, S., 1998. Bending, buckling and free vibration of laminated composite with a transverse shear stress continuity model. Compos. Part B 29B, 223–234.

Karama, M., Afaq, K.S., Mistou, S., 2003. Mechanical behaviour of laminated composite beam by the new multi-layered laminated composite structures model with transverse shear stress continuity. Int. J. Solids Struct. 40, 1525–1546.

Lekhnitskii, S.G., 1935. Strength calculation of composite beams. Vestn. Inzh. Tekh. 9.

Levy, M., 1877. Memoire sur la theorie des plaques elastique planes. J. Math. Pures Appl. 30, 219–306.

Lo, K.H., Christensen, R.M., Wu, E.M., 1977. A high-order theory of plate deformation – Part 2: Laminated plates. J. Appl. Mech. 44 (4), 669–676.

Lucintel, 2013. Opportunities for composites in European automotive market 2013– 2018, Research report, Lucintel – Global Management Consulting & Market Research Firm.

Mantari, J.L., Oktem, A.S., Guedes Soares, C., 2011. Static and dynamic analysis of laminated composite and sandwich plates and shells by using a new higherorder shear deformation theory. Compos. Struct. 94, 37–49.

Mindlin, R.D., 1951. Influence of rotary inertia and shear on flexural motion of isotropic elastic plates. ASME J. Appl. Mech. 18, 31–38.

Murakami, H., 1986. Laminated composite plate theory with improved in-plane responses. ASME J. Appl. Mech. 53, 661–666.

Neves, A.M.A., Ferreira, A.J.M., Carrera, E., Cinefra, M., Roque, C.M.C., Jorge, R.M.N., 2013. Free vibration analysis of functionally graded shells by a higher-order shear deformation theory and radial basis functions collocation, accounting for through-the-thickness deformations. Eur. J. Mech A – Solid 37, 24–34.

Pagano, N.J., 1969. Exact solutions for composite laminates in cylindrical bending. J. Compos. Mater. 3 (3), 398–411.

Pagano, N.J., 1970a. Exact solutions for rectangular bidirectional composites and sandwich plates. J. Compos. Mater. 4 (20), 20–34.

Pagano, N.J., 1970b. Influence of shear coupling in cylindrical bending of anisotropic laminates. J. Compos. Mater. 4 (3), 330–343.

Reddy, J.N., 1983. A refined nonlinear theory of plates with transverse shear deformation. Int. J. Solids Struct. 20 (9), 881–896.

Reddy, J.N., 1986. A refined shear deformation theory for the analysis of laminated plates, Contractor Report 3955, National Aeronautics and Space Administration.

Reissner, E., 1944. On the theory of bending of elastic plates. J. Math. Phys. 23, 184– 191.

Reissner, E., 1945. The effect of transverse shear deformation on the bending of elastic plates. J. Appl. Mech. 12 (30), A69–A77.

Reissner, E., 1975. On transverse bending of plates, including the effect of transverse shear deformation. Int. J. Solids Struct. 11, 569–573.

Reissner, E., 1984. On a certain mixed variational theorem and a proposed application. Int. J. Numer. Methods Eng. 20 (7), 1366–1368.

Ren, J.G., 1986a. A new theory of laminated plates. Compos. Sci. Technol. 26, 225– 239.

Ren, J.G., 1986b. Bending theory of laminated plates. Compos. Sci. Technol. 27, 225– 248.

Soldatos, K.P., 1992. A transverse shear deformation theory for homogeneous monoclinic plates. Acta. Mech. 94, 195–220.

Stein, M., 1986. Nonlinear theory for plates and shells including the effect of transverse shearing. AIAA J. 24, 1537–1544.

Tessler, A., 1993. An improved theory of 1,2 order for thick composite laminates. Int. J. Solids Struct. 30 (7), 981–1000.

Tessler, A., Di Sciuva, M., Gherlone, M., 2007. Refinement of Timoshenko beam theory for composite and sandwich beams using zigzag kinematics, Technical Publication 215086, National Aeronautics and Space Administration.

Tessler, A., Di Sciuva, M., Gherlone, M., 2009. Refined zigzag theory for laminated composite and sandwich plates. Technical Publication 215561, National Aeronautics and Space Administration.

Tessler, A., Di Sciuva, M., Gherlone, M., 2010. Refined zigzag theory for homogeneous, laminated composite, and sandwich plates: a homogeneous limit methodology for zigzag function selection, Technical Publication 216214, National Aeronautics and Space Administration.

Tessler, A., Di Sciuva, M., Gherlone, M., 2010b. A consistent refinement of first-order shear deformation theory for laminated composite and sandwich plates using improved zigzag kinematics. J. Mech. Mater. Struct. 5 (2), 341–367.

Timoshenko, S., 1934. Theory of Elasticity. McGraw-Hill Book Company Inc., New York.

Toledano, A., Murakami, H., 1987. A composite plate theory for arbitrary laminate configurations. J. Appl. Mech. 54 (1), 181–189.

Touratier, M., 1991. An efficient standard plate theory. Int. J. Eng. Sci. 29, 901–916. Vlasov, B.F., 1957. On the equations of bending of plates. Dokla. Ak. Nauk. Azerbeijanskoi-SSR 3, 955–979.

Whitney, J.M., 1969. The effect of transverse shear deformation on the bending of laminated plates. J. Compos. Mater. 3, 534–547.

Whitney, J.M., 1972. Stress analysis of thick laminated composite and sandwich plates. J. Compos. Mater. 6 (4), 426–440.

Whitney, J.M., Pagano, N.J., 1970. Shear deformation in heterogeneous anisotropic plates. J. Appl. Mech. 37, 1031–1036.

Williams, T.O., 2001. Efficiency and accuracy considerations in a unified plate theory with delamination. Compos. Struct. 52, 27–40.

Williams, T.O., 2005. A generalized, multilength scale framework for thermo diffusional-mechanically coupled, nonlinear, laminated plate theories with delaminations. Int. J. Solids Struct. 42 (5–6), 1465–1490.

Williams, T.O., 2008. A new theoretical framework for the formulation of general, nonlinear, multiscale plate theories. Int. J. Solids Struct. 45 (9), 2534–2560.

Yang, P.C., Norris, C.H., Stavsky, Y., 1966. Elastic wave propagation in heterogeneous plates. Int. J. Solids Struct. 2, 665–684

CITATION

R.M.J. Groh, P.M. Weaver, On displacement-based and mixed-variational equivalent single layer theories for modelling highly heterogeneous laminated beams, International Journal of Solids and Structures, Volume 59, 1 May 2015, Pages 147-170, ISSN 0020-7683, http://dx.doi.org/10.1016/j.ijsolstr.2015.01.020.

CHAPTER 6

An Extended Finite Element Model for Modelling Localised Fracture of Reinforced Concrete Beams in Fire

Feiyu Liao, Zhaohui Huang

Department of Mechanical, Aerospace and Civil Engineering, College of Engineering, Design and Physical Sciences, Brunel University, Uxbridge, Middlesex UB8 3PH, UK

ABSTRACT

A robust finite element procedure for modelling the localised fracture of reinforced concrete beams at elevated temperatures is developed. In this model a reinforced concrete beam is represented as an assembly of 4-node quadrilateral plain concrete, 3-node main reinforcing steel bar, and 2-node bond-link elements. The concrete element is subdivided into layers for considering the temperature distribution over the cross-section of a beam. An extended finite element method (XFEM) has been incorporated into the concrete elements in order to capture the localised cracks within the concrete. The model has been validated against previous fire test results on the concrete beams.

INTRODUCTION

Localised fracture of reinforced concrete members has recently been of interest to many researchers and engineers. Under fire conditions, reinforced concrete structural members (such as beams or slabs) are often forced into high deformation. This results in the formation of large individual cracks within the members, which has been observed in previous experimental tests [1], [2] and [3]. These large individual cracks influence the exposure condition of the reinforcing steel bar to the fire. In some cases the steel reinforcements are directly exposed to fire, whereby significantly reducing the fire resistance of the structures. In some extreme cases, localised large cracks could even result in integrity failure of the structures [1], [2] and [3]. A key factor in assessing the fire resistance of the structures is through predicting the localised fracture of their structural members. Recently, the performance-based approach has been used in the fire safety design of reinforced concrete structures, which requires the use of accurate numerical models for predicting the response of structural members in fire. In the past two decades, plenty of numerical simulations and analyses have been conducted for modelling concrete structures at elevated

temperatures [4], [5], [6], [7], [8], [9], [10], [11], [12],[13] and [14]. Those studies were all based on the continuum approach, in which smeared cracking was adopted to simulate the cracks within concrete members. Existing research indicates that models based on smeared cracking can predict global responses, such as deflection and structural stability, with reasonable accuracy. However, the smeared cracking model cannot capture the localised fracture within structural members, and quantitatively predict crack openings. As far as performance-based fire safety design is concerned, predicting the opening of individual cracks at critical sections of critical members can be a crucial issue when evaluating the reliability of structures under fire conditions. Little research has yet been done on modelling localised fractures for reinforced concrete structural members under fire conditions.

In the past, a discrete-cracking model has been used successfully for modelling the formation and propagation of cracks in structural members, when the crack path is known in advance. However, this approach has to limit the cracks to inter-element boundaries, which might cause mesh bias, or requires performing costly re-meshing during the analysis process. To model individual cracks more effectively, the extended finite element method (XFEM) was introduced [15] and [16], based on the partition of unity theory [17]. The XFEM approaches in conjunction with cohesive-zone models [18],[19], [20] and [21] allow displacement jumps within conventional finite elements to analyse localisation and fracture in engineering materials. In the last decade, the XFEM has been successfully extended to many applications, such as multiple cracks in brittle materials, intersecting cracks and dynamic crack growth [22], [23], [24], [25] and [26]. In terms of computer implementation of enriched finite element methods, a general structure for an object-oriented enriched finite element code (the XFEM library) was presented by [27], which had been designed to meet all natural requirements for modularity, extensibility, and robustness. Another open-source software framework called PERMIX for multiscale modelling of material failure was presented by [28]. The integration method for the XFEM based on Schwarz–Christoffel mappings was proposed by [29] to simplify the numerical integration on arbitrary polygonal domains. The application of strain smoothing in finite elements was extended to the extended finite element method to form the smoothed extended finite element method (cell-based smoothed XFEM, edge-based smoothed XFEM, and node-based smoothed XFEM) [30],[31] and [32]. By transforming interior integration into boundary integration, strain smoothing simplifies the integration of discontinuous approximations of the XFEM and suppresses the need to integrate singular functions numerically. The smoothed XFEM is insensitive to mesh distortion and locking and could be a competitive alternative to solve complex 3D problems. The strain smoothing method was extended to higher-order elements by [33], and it also concluded that the method is only beneficial when the enrichment functions are polynomial. Besides the XFEM, there are also some alternate approaches for modelling the strong discontinuity. The numerical results of the embedded finite element method (EFEM) and XFEM were compared in [34], and various methods for numerical modelling of multifield fracturing, such as interface and embedded discontinuity elements, XFEM, thick level set and phase field models, and a discrete crack approach with adaptive remeshing, were discussed in [35]. Recently, the meshfree method based on a partition of unity concept was also developed to model concrete and more general non-linear materials [36], [37] and [38]. This method has been used to successfully model the reinforced concrete structural members at the ambient temperature [39] and [40], where a coupled particle-finite element approach was adopted and the reinforcement was coupled with the concrete via a 'barscale' bond model for modelling the pullout and splitting failure. However, so far, limited efforts have been made to use the XFEM in modelling reinforced concrete structural members in fire.

The main objective of this paper is to develop a robust finite element procedure for modelling the localised fracture of reinforced concrete members in fire conditions. The model developed can be used for structural fire engineering design of reinforced concrete beams and enable engineers to assess both the structural stability (global response) and the integrity (localised fracture) of the beams. In the past, the majority of reinforced concrete beams at elevated temperatures have been simulated by the conventional finite element method, in which the generalised isoparametric elements have usually been used. In the procedure proposed in this paper the isoparametric elements are still employed so that relatively small modifications of the available finite element model are required. The new procedure could be easily applied to the fire design for practical building structures. In this paper a 2D model is used to model reinforced concrete beams. Since mesh distortion and locking are not the main concerns within the scope of this paper, only the straight crack is considered and the standard XFEM formulations and numerical integration procedure are employed. In this new model a reinforced concrete beam is represented as an assembly of 4-node quadrilateral plain concrete, 3-node main reinforcing steel bar, and 2-node bond-link elements. The extended finite element method (XFEM) is incorporated into plain concrete elements in order to capture the localised cracks of concrete within the member. The original contributions of the model presented in this paper are:

- Combine the XFEM plain concrete element with the reinforcing bar element and bond-link element successfully. Due to the bond-link element and plain concrete element sharing the same node, one important issue with which the model should deal is the compatibility of nodal displacements referenced to the plain concrete element and bond-link element. This is due to the nodal displacement of the cracked XFEM plain concrete element being divided into two parts: continuous part and discontinuous part. These displacements are not compatible with the nodal displacement of the bond-link element. Therefore, a special shifted enhancement function is used in order to obtain the total nodal displacement (including both continuous and discontinuous parts) of the cracked XFEM plain concrete elements. This satisfies the compatibility of the nodal displacements of both the XFEM plain concrete element and bond-link element.

- With the help of the bond-link element and steel bar element, the developed model has the capability to consider the influence of the bond characteristic between the concrete and reinforcing steel bar on the initiation and propagation of each individual crack within the reinforced concrete beam. Due to the influence of the reinforcing steel bar, the Newton–Raphson iteration procedure can be employed to solve this very nonlinear problem up to the failure of the whole beam. This is significantly different with conventional XFEM models, in which a complex solution procedure needs to be developed.

- Even for the adoption of a 2D model for modelling plain concrete, the model developed in this paper is still complex, because the effects of temperatures induced by fire need to be taken into account. The XFEM plain concrete elements are subdivided into layers for considering the temperature distribution over the cross-section of a beam. Since the temperature varies across different layers, a robust criterion has been developed to determine the initiation of individual cracks within the XFEM plain concrete elements. Moreover, the complications of structural behaviour in fire, such as thermal expansion, degradation of bond characteristics between a reinforcing steel bar and concrete, and the change of material properties with temperature, are modelled.

The new model has been validated against some previous fire tests of reinforced concrete beams. It is clear that the developed nonlinear procedure proposed in this paper can predict cracking patterns (flexural cracks and shear cracks) of the reinforced concrete beams properly. The model is capable of predicting the global response of reinforced concrete beams in fire with good accuracy and, at the same time, capturing the formation and propagation of individual localised cracks within the beams. The influences of bond characteristics between the concrete and reinforcing steel bar on the deflection and crack opening are also examined in this paper. The model presented in this paper provides a very useful tool for researchers and designers to assess the integrity of reinforced concrete structural members under fire conditions.

DEVELOPMENT OF THE NONLINEAR PROCEDURE

As shown in Fig. 1, a reinforced concrete beam is modelled as an assembly of plain concrete, main reinforcing steel bar, and bond-link elements. The plain concrete elements are subdivided into layers to take into account the temperature distribution over the cross-section of a beam. The bond-link elements are used to represent the interaction between the plain concrete and reinforcing steel bar elements.

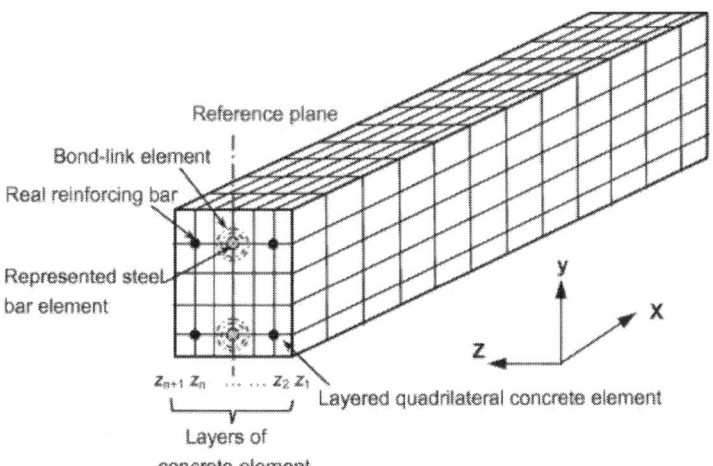

Figure 1. Nonlinear layered finite element procedure for modelling a reinforced concrete beam.

Layered Quadrilateral Concrete Elements with Extended Finite Element Formulations

Fig. 2 shows a 4-node layered quadrilateral element to simulate the plain concrete for a reinforced concrete beam under fire conditions. Each node of the element contains two translational degrees of freedom. In order to consider the temperature distribution over the beam cross-section, the plain concrete elements are divided into layers in the z direction, and each layer can have a different but uniform temperature. The initial material properties of each layer may be different. Concrete layers are in a state of plane stress, so the material property, temperature, load and deformation of the element are symmetric to the reference plane (mid-plane, as shown in Fig. 2) along the thickness [41]. Within the element the stress–strain relationships may change independently for each layer. Since temperature-dependent constitutive models are used in this study, the material properties and thermal expansions vary at different layers, but are constant within a layer at each temperature step.

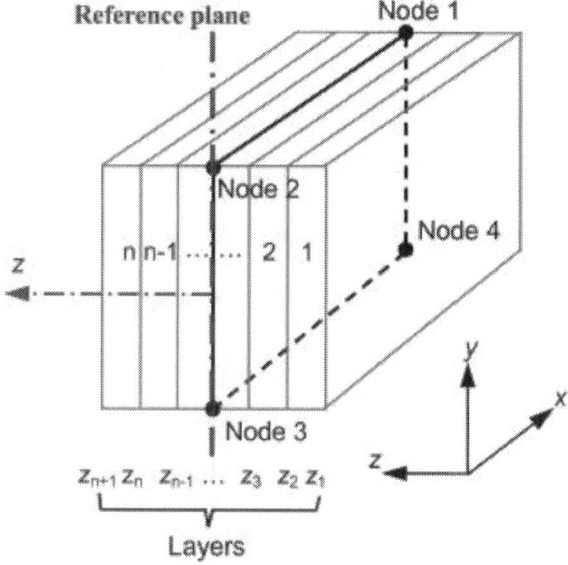

Figure 2. A layered 4-node quadrilateral plain concrete element.

Element Stiffness Matrix, K

In order to model the localised fracture of plain concrete, the extended finite element method (XFEM) is used for the development of plain concrete elements. The model proposed here can capture individual concrete cracks without remeshing and calculating the magnitude of individual crack openings during the analysis. The key idea of the extended finite element method

(XFEM) is to use the partition of unity for describing the discontinuous displacement, and then the displacement field is approximated by the sum of the regular displacement and the enhancement displacement fields [16]. In order to do this, extra degrees of freedom are added on the enhanced nodes to represent the enhancement displacement field. Special enhancement functions are also employed to realise the displacement jump over the discontinuity. Note that many applications of the XFEM have been conducted using a step function to enrich the element which is completely cut by a crack and using branch functions to enrich the element which the crack tip located inside the element. In this paper, for simplicity, it is assumed that the crack tip is always located on an edge of an element; thus, the cracked element can be successfully enriched by the *sign* function only without other enrichment functions. Thus, the crack branching is not included in the proposed procedure, assuming that a particular element contains one crack only. The main purpose of the model developed in this paper is to predict the global response of reinforced concrete beams in fire with reasonable accuracy and, at the same time, the major localised cracks within the beam can also be captured. Therefore, in order to enhance the computational efficiency of the proposed model, precise modelling of the crack tip and crack branching is not considered in this paper.

Considering a four-node quadrilateral element crossed by a discontinuity (Γ_d) (see Fig. 3), the domain is divided into two distinct domains referenced to an element, which are represented as Ω^+ and Ω^- on the different sides of the discontinuity in an element. Figs. 3(a) and (b) give the definition of sub-domains Ω^+ and Ω^- where a discontinuity cuts a quadrilateral element in two different possible ways, respectively. Then, the total displacement field $\mathbf{u}$ consists of a continuous regular displacement field $\mathbf{u}_{cont}$ and a discontinuous displacement field $\mathbf{u}_{dis}$ [18], that is:

$$\mathbf{u} = \mathbf{u}_{cont} + \mathbf{u}_{dis} = \sum_1^4 N_i \mathbf{u}_i + \sum_1^4 N_i \Psi_i(\mathbf{x}) \mathbf{a}_i$$

(1)

where N_i is the shape function, $\mathbf{u}_i$ is the regular node displacement, $\mathbf{a}_i$ is the additional node displacement to describe the discontinuity, and $\Psi_i(\mathbf{x})$ is the enhancement function:

$$\Psi_i(\mathbf{x}) = sign(x) - sign(x_i) \qquad (i = 1 \sim 4)$$

(2)

in which *sign* is the sign function and defined as:

$$sign(x) = \begin{cases} +1 & \text{if } x \in \Omega^+ \\ -1 & \text{if } x \in \Omega^- \end{cases}$$

(3)

Note that the *sign* function enrichment is equivalent to the Heaviside step function enrichment which has been used in many previous XFEM models. But the *sign* function appears more symmetrical [19]. $sign(x_i)$ is the value of the *sign* function of the *i-th* node in a quadrilateral element. Taking the quadrilateral element in Fig. 3(a) as an example: $sign(x_2)=sign(x_3)=-1$ for nodes 2 and 3 and $sign(x_1)=sign(x_4)=+1$ for nodes 1 and 4, respectively. Compared with the conventional XFEM models, the *sign* function given in Eq. (2) is shifted by $sign(x_i)$. According to [19], using the shifted *sign* function can make the enrichment displacement field vanish outside the enhanced element.

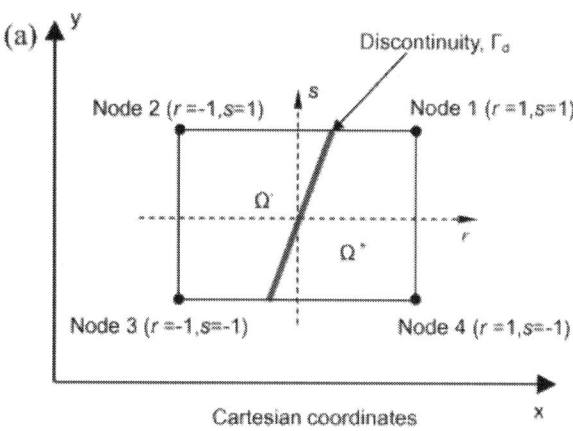

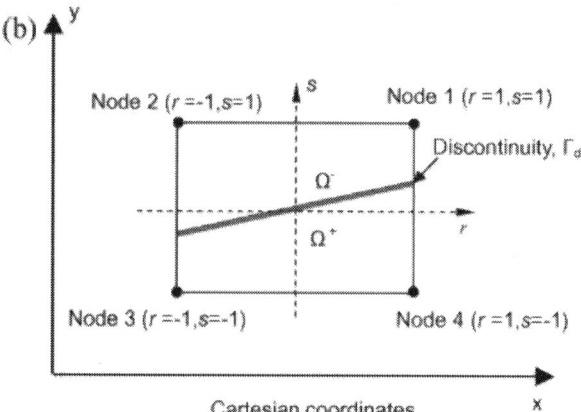

Figure 3. A 4-node quadrilateral element crossed by a discontinuity Γ_d.

One significant advantage of using the *sign* function is that only the elements cut by the crack need to be enhanced, as the resulting enhancement vanishes in all elements not crossed by the crack. The utilisation of a shifted *sign* function may greatly simplify the implementation of the extended finite element model, without altering the approximating basis. Especially for modelling reinforced concrete structures, this advantage is more significant because multiple cracks normally are distributed within a reinforced concrete member (due to the bond action between steel bars and concrete). Other than the simplification in terms of implementing the procedure, the key reason for using the shifted enhancement function is to obtain the total nodal displacement directly from the procedure, rather than only the regular part of XFEM nodal displacement being outputted from the procedure [42]. This makes the compatibility of total nodal displacements of the plain concrete element and bond-link element feasible. Therefore, the bond-link element can be used to link plain concrete elements and steel bar elements in a conventional way, such as through the continuous approach, for modelling localised cracking within a reinforced concrete member.

In the case of the four-node quadrilateral element (Fig. 3), the element nodal displacement vector $\hat{\mathbf{u}}$ can be represented as:

$$\hat{\mathbf{u}} = \left\{ \begin{matrix} \mathbf{u}_i \\ \mathbf{a}_i \end{matrix} \right\} = \begin{bmatrix} u_1 & v_1 & u_2 & v_2 & u_3 & v_3 & u_4 & v_4 & a_1 & b_1 & a_2 & b_2 & a_3 & b_3 & a_4 & b_4 \end{bmatrix}^T$$

(4)

where u_i and v_i are the regular nodal displacements related to x and y coordinates, respectively, and a_i and b_i are the enhanced nodal displacements related to x and y coordinates, respectively.

Thus, the strains (ε) within an enhanced element consist of the regular and enhancement parts, which are related to the regular nodal and enhanced nodal displacements respectively. The strain vector ε can be expressed as:

$$\varepsilon = \varepsilon_{cont} + \varepsilon_{dis} = \left\{ \begin{matrix} \varepsilon_x \\ \varepsilon_y \\ \gamma_{xy} \end{matrix} \right\} = \mathbf{B}\hat{\mathbf{u}} = \begin{bmatrix} \mathbf{B}^u_{sta} & \mathbf{B}^a_{enr} \end{bmatrix} \left\{ \begin{matrix} \mathbf{u}_i \\ \mathbf{a}_i \end{matrix} \right\}$$

(5)

where $\varepsilon cont$ is the continuous strain, and εdis is the discontinuous strain. $\mathbf{B}^u_{sta}$ is the standard strain–displacement transformation matrix corresponding to the regular degrees of freedom $\mathbf{u}_i$, and $\mathbf{B}_{enr}^{a}$ is the enrichment strain–displacement transformation matrix corresponding to the additional degrees of freedom $\mathbf{a}_i$.

The strain–displacement transformation matrix $\mathbf{B}$ including the regular part and enhancement part can be obtained as $\mathbf{B} = \begin{bmatrix} \mathbf{B}_{sta}^{u} \mathbf{B}_{enr}^{a} \end{bmatrix}$, in which:

$$
\mathbf{B}_{sta}^{u} = \mathbf{LN} = \mathbf{L}\begin{bmatrix} N_1 & 0 & N_2 & 0 & N_3 & 0 & N_4 & 0 \\ 0 & N_1 & 0 & N_2 & 0 & N_3 & 0 & N_4 \end{bmatrix}
$$

$$
= \begin{bmatrix} B_{sta1x} & 0 & B_{sta2x} & 0 & B_{sta3x} & 0 & B_{sta4x} & 0 \\ 0 & B_{sta1y} & 0 & B_{sta2y} & 0 & B_{sta3y} & 0 & B_{sta4y} \\ B_{sta1y} & B_{sta1x} & B_{sta2y} & B_{sta2x} & B_{sta3y} & B_{sta3x} & B_{sta4y} & B_{sta4x} \end{bmatrix} \quad (6)
$$

$$
\mathbf{B}_{enr}^{a} = \Psi_i(\mathbf{x})\mathbf{LN} = \mathbf{L}\begin{bmatrix} \Psi_1(x)N_1 & 0 & \Psi_2(x)N_2 & 0 & \Psi_3(x)N_3 & 0 & \Psi_4(x)N_4 & 0 \\ 0 & \Psi_1(x)N_1 & 0 & \Psi_2(x)N_2 & 0 & \Psi_3(x)N_3 & 0 & \Psi_4(x)N_4 \end{bmatrix}
$$

$$
= \begin{bmatrix} B_{enr1x} & 0 & B_{enr2x} & 0 & B_{enr3x} & 0 & B_{enr4x} & 0 \\ 0 & B_{enr1y} & 0 & B_{enr2y} & 0 & B_{enr3y} & 0 & B_{enr4y} \\ B_{enr1y} & B_{enr1x} & B_{enr2y} & B_{enr2x} & B_{enr3y} & B_{enr1x} & B_{enr4y} & B_{enr4x} \end{bmatrix}
$$

$$(7)$$

where $\mathbf{N}$ is the shape function of a general quadrilateral element [43], $\Psi_i(\mathbf{x})(i = 1\text{–}4)$ is the enrichment function given in Eq. (2), and the matrix $\mathbf{L}$ contains differential operators.

Since the effect of thermal expansion is included in the model, the total strains (ε) include both thermal and stress-related strains at elevated temperatures. The stress-related strains can be obtained by deducing the thermal strains (ε_T) from the total strains (ε). If strains are reasonably small the stress vector σ can be obtained from the stress-related strain vector as:

$$
\sigma = \begin{Bmatrix} \sigma_x \\ \sigma_y \\ \tau_{xy} \end{Bmatrix} = \mathbf{D}(\varepsilon - \varepsilon_T) = \mathbf{D}\big(\mathbf{B}_{sta}^{u}\mathbf{u}_i + \Psi_i(\mathbf{x})\mathbf{B}_{sta}^{u}\mathbf{a}_i - \varepsilon_T\big)
$$

$$(8)$$

in which $\mathbf{D}$ is the constitutive matrix of concrete related to plane stress.
In a finite element model, the equilibrium conditions between internal and external 'forces' have to be satisfied. To form the element stiffness matrix

and internal force vector, the virtual work equation without body force reads as:

$$\mathbf{f}^{int} = \int_{\Omega} \mathbf{B}^T \sigma \, d\Omega = \mathbf{f}^{ext} \tag{9}$$

where $\mathbf{f}int$ is the internal force vector, and $\mathbf{f}^{ext}$ is the external force vector. In this study the cracked concrete is treated as a quasi-brittle heterogeneous material, and the cohesive crack concept is used for simulating quasi-brittle fracture. The internal force vector $\mathbf{f}int$ contains the regular part $(\mathbf{f}_u^{int})$, the enhancement part $(\mathbf{f}_a^{int})$, and the traction part $(\mathbf{f}_\Gamma^{int})$. The regular internal force $(\mathbf{f}_u^{int})$ balances the external force $(\mathbf{f}^{ext})$, and the enhancement part $(\mathbf{f}_a^{int})$ is related to the traction of the crack $(\mathbf{f}_\Gamma^{int})$ only [18], that is:

$$\mathbf{f}_u^{int} = \int_{\Omega} \mathbf{B}_{sta}^{u^T} \sigma \, d\Omega = \mathbf{f}^{ext} \tag{10}$$

$$\mathbf{f}_a^{int} + \mathbf{f}_\Gamma^{int} = \int_{\Omega^+,\Omega^-} \mathbf{B}_{enr}^{a^T} \sigma \, d\Omega + \int_{\Gamma_d} \overline{\mathbf{N}}^T \mathbf{t}_a \, d\Gamma_d = 0 \tag{11}$$

where $\mathbf{t}_a$ is the traction acting on the discontinuity and can be written as:

$$\mathbf{t}_a = \begin{Bmatrix} t_{an} \\ t_{as} \end{Bmatrix} = \mathbf{T}_a \mathbf{w} = \begin{bmatrix} T_{an} & 0 \\ 0 & 0 \end{bmatrix} \begin{Bmatrix} w_n \\ w_s \end{Bmatrix} \tag{12}$$

where t_{an} and t_{as} are the traction normal and tangential to a crack, respectively; w_n and w_s are the crack opening normal and tangential to a crack, respectively; and T_{an} is the tangent stiffness of the traction–separation law.

In order to solve the nonlinear problem, an incremental solution procedure needs to be developed. By substituting the rate form of the constitutive relations of Eq. (8) into Eqs.(10) and (11), the element stiffness matrix in terms of incremental displacements can be obtained as:

$$\mathbf{K} \, d\hat{\mathbf{u}} = \begin{bmatrix} \mathbf{K}_{uu} & \mathbf{K}_{ua} \\ \mathbf{K}_{au} & (\mathbf{K}_{aa} + \mathbf{K}_\Gamma) \end{bmatrix} \begin{Bmatrix} d\mathbf{u}_i \\ d\mathbf{a}_i \end{Bmatrix} = \begin{Bmatrix} \mathbf{f}^{ext} \\ 0 \end{Bmatrix} - \begin{Bmatrix} \mathbf{f}_u^{int} \\ \mathbf{f}_a^{int} + \mathbf{f}_\Gamma^{int} \end{Bmatrix} \tag{13}$$

where $\mathbf{K}uu$ is the element stiffness matrix referenced to the regular degrees of freedom, $\mathbf{K}aa$ is the element stiffness matrix related to the enhancement degrees of freedom, $\mathbf{K}_{ua} = \mathbf{K}_{au}^T$ is related to both, and $\mathbf{K}_\Gamma$ is the element stiffness matrix related to traction. They are expressed as:

$$\mathbf{K}_{uu} = \int_\Omega \mathbf{B}_{sta}^{u^T} \mathbf{D} \mathbf{B}_{sta}^u d\Omega = \iint_A \mathbf{B}_{sta}^{u^T} \left(\int \mathbf{D}_l dz \right) \mathbf{B}_{sta}^u \, dx \, dy \tag{14}$$

$$\mathbf{K}_{ua} = \int_{\Omega^+ , \Omega^-} \mathbf{B}_{sta}^{u^T} \mathbf{D} \mathbf{B}_{enr}^a \, d\Omega = \iint_{A^+ , A^-} \mathbf{B}_{sta}^{u^T} \left(\int \mathbf{D}_l dz \right) \mathbf{B}_{enr}^a \, dx \, dy \tag{15}$$

$$\mathbf{K}_{au} = \int_{\Omega^+ , \Omega^-} \mathbf{B}_{enr}^{a^T} \mathbf{D} \, \mathbf{B}_{sta}^u \, d\Omega = \iint_{A^+ , A^-} \mathbf{B}_{enr}^{a^T} \left(\int \mathbf{D}_l dz \right) \mathbf{B}_{sta}^u \, dx \, dy \tag{16}$$

$$\mathbf{K}_{aa} = \int_{\Omega^+ , \Omega^-} \mathbf{B}_{enr}^{a^T} \mathbf{D} \mathbf{B}_{enr}^a \, d\Omega = \iint_{A^+ , A^-} \mathbf{B}_{enr}^{a^T} \left(\int \mathbf{D}_l dz \right) \mathbf{B}_{enr}^a \, dx \, dy \tag{17}$$

$$\mathbf{K}_\Gamma = \int_{\Gamma_d} \bar{\mathbf{N}}^T \mathbf{T}_a \bar{\mathbf{N}} d\Gamma_d - \int_{\Gamma_d} \bar{\mathbf{N}}^T \mathbf{O}^T \mathbf{T}_d \mathbf{O} \bar{\mathbf{N}} d\Gamma_d$$
$$= \int_{\Gamma_d} \bar{\mathbf{N}}^T \mathbf{O}^T \left(\int \mathbf{T}_{dl} dz \right) \mathbf{O} \bar{\mathbf{N}} d\Gamma_d \tag{18}$$

where $\bar{\mathbf{N}} = 2(\mathbf{N})$ and $\mathbf{N}$ is the shape function, $\mathbf{T}_d$ is the tangent stiffness of the traction–separation law, and $\mathbf{O}$ is the orthogonal transformation matrix – for the transformation of the local orientation of the discontinuity to the global coordinate system. The expression of $\mathbf{O}$ can be found in [44].

In this study, Gauss quadrature is employed to calculate the stiffness matrix of quadrilateral elements. Therefore, all stresses, strains, and the constitutive matrix of material discussed above correspond to Gauss integration points. Since the elements are divided into layers along the z-axis (see Fig. 2), and the material properties are assumed to be constant within each layer at each time or temperature step, the matrices $\mathbf{D}$ and $\mathbf{T}_d$ in Eqs. (14), (15), (16), (17) and (18) at a Gauss point are a function of z only, and the inner integrations $\left(\int \mathbf{D}_l dz \right)$ and $\left(\int \mathbf{T}_{dl} dz \right)$ can be performed separately. Integration along the z- axis is replaced by summation over the layers as

$$\int \mathbf{D}_l dz = \sum_{l=1}^{n} (z_{l+1} - z_l) \mathbf{D}_l \tag{19}$$

$$\int \mathbf{T}_{dl}\,dz = \sum_{l=1}^{n}(z_{l+1} - z_{l})\mathbf{T}_{dl}$$

(20)

where

z_l is the distance from the reference plane to the l-th layer (Fig. 2),

$\mathbf{D}_l$ is the material stiffness matrix for the l-th layer,

$\mathbf{T}_{dl}$ is the tangent stiffness matrix of traction–separation law for the l-th layer, and

n is the total number of element layers.

Element Internal Force Vector, f int
Using the principle of virtual work, the internal force vectors can be written as [18]:

$$\mathbf{f}_{u}^{int} = \int_{\Omega} \mathbf{B}_{sta}^{u'} \,\sigma\,d\Omega = \iint_{A} \mathbf{B}_{sta}^{u'}\left(\int \sigma_l\,dz\right)dx\,dy$$

(21)

$$\mathbf{f}_{a}^{int} = \int_{\Omega'.\Omega} \mathbf{B}_{ent}^{a'} \,\sigma\,d\Omega = \iint_{A'.A} \mathbf{B}_{ent}^{a'}\left(\int \sigma_l\,dz\right)dx\,dy$$

(22)

$$\mathbf{f}_{l}^{int} = \int_{\Gamma_d} \mathbf{\bar{N}}^{T}\mathbf{t}_a\,d\Gamma_d = \int_{\Gamma_d} \mathbf{\bar{N}}^{T}\mathbf{O}^{T}\left(\int \mathbf{t}_{dl}\,dz\right)d\Gamma_d$$

(23)

As mentioned above, the Gauss quadrature is used to calculate $\mathbf{f}_{u}^{int}$, $\mathbf{f}_{a}^{int}$ and $\mathbf{f}_{l}^{int}$. Therefore, $(\int \sigma\,dz)$ and $(\int \mathbf{t}_{dl}\,dz)$ at a Gauss point are a function of z only, and the inner integrations $\int \sigma_l\,dz$ and $\int \mathbf{t}_{dl}\,dz$ in Eqs. (21), (22) and (23) can be expressed by summation over the layers as:

$$\int \sigma_l\,dz = \sum_{l=1}^{n}(z_{l+1} - z_l)\sigma_l$$

(24)

$$\int \mathbf{t}_{dl} \, dz = \sum_{l=1}^{n} (z_{l+1} - z_l) \mathbf{t}_{dl}$$

(25)

where σ_1 is the stress vector in the l-th layer, $\mathbf{t} dl$ is the traction in the l-th layer, and n is the total number of layers.

In this study, Gauss quadrature is employed to calculate the stiffness matrix and internal force vector of the concrete element. $\mathbf{K}uu$ and $\mathbf{f}_u^{int}$ can be integrated in the usual way over the whole domain (Ω), but for $\mathbf{K}_{aa}$, $\mathbf{K}_{ua}$, $\mathbf{K}_{au}$ and $\mathbf{f}_a^{int}$, integration should be performed separately on both sides $(\Omega^+$ and $\Omega^-)$ of the crack, respectively. This means that the *sign* function *sign* (x) needed to be applied for each Gauss point within the element. In the current model, a crack is represented by a straight line within the enhancement element, with two Gauss points employed to integrate the discontinuity terms $\mathbf{K}_\Gamma$ and $\mathbf{f}_\Gamma^{int}$ over the discontinuity Γ_d using a one-dimensional integration scheme. For the regular four-node element without a crack, four Gauss integration points are used (as recommended by Bathe [43]). But for those enhanced elements containing a crack, conventional four Gauss integration is insufficient to distinguish the enhancement function from a constant function over different sides $(\Omega^+$ and $\Omega^-)$ of the crack, resulting in linearly independent shape functions [18]. Therefore, the enhanced elements need to be integrated separately on each side of the crack. Note that there are many different integration schemes that could be used for an enhanced quadrilateral element with sufficient accuracy [45]. In this paper the scheme that partitioned the element into sub-triangles is adopted for its flexibility because it might be desirable to add other enhancement functions into the procedure in future research. Then, in the case of the element being cut by a crack into two sub-quadrilaterals, as shown in Fig. 4(a), four sub-triangles with 12 Gauss points are applied within each sub-quadrilateral. Fig. 4(b) shows the element being cut by a crack into a pentagon and a triangle. In this case, five sub-triangles with 15 Gauss points are applied within the pentagon, and three sub-triangles with nine Gauss points are used within the triangle. More detailed information related to the integration schemes can be found in [44]. Due to the high non-linearity of the current model, a full Newton–Raphson solution procedure is adopted.

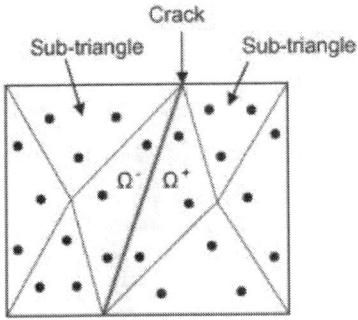

• Integration point

(a) A crack cutting a concrete element
into two quadrilaterals

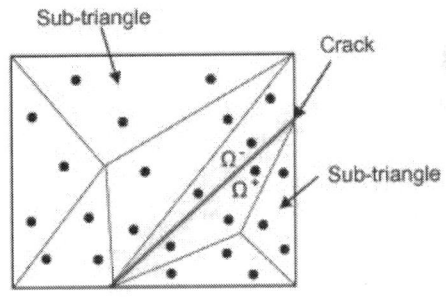

• Integration point

(b) A crack cutting a concrete element into
a pentagon and a triangle

Figure 4. Integration scheme for an enhanced 4-node quadrilateral element crossed by a crack.

Constitutive Modelling of Concrete at Elevated Temperatures

Before cracking or crushing occurs, the concrete is assumed to be isotropic, homogeneous, and linearly elastic. Barzegar-Jamshidi [46] proposed a biaxial concrete failure envelope at an ambient temperature, which was based on a slight modification of the Kupfer and Gerstle [47] expressions. At present, there are still very little data and few theoretical models available regarding the constitutive modelling of concrete under biaxial states of stress at elevated temperatures. Based on the Barzegar-Jamshidi [46]model, Huang et al. [11] developed a biaxial concrete failure envelope at elevated temperatures by considering all of the relevant material properties as temperature-dependent. As shown in Fig. 5, with the increasing temperatures the area enclosed by the failure envelope tends to be decreasing. The model was validated against the test results in [11] and [12]. Therefore, this model is adopted to determine the cracking and crushing of concrete in this paper. In the figure, $f'_c(T)$ and $f'_t(T)$ are the temperature-dependent compressive strength and tensile strength of concrete, respectively; σ_{c1} and σ_{c2} are the principal stresses. The failure surfaces of the biaxial strength envelope are divided into four regions which depend on the stress state as represented by the principal stress ratio $\alpha = \sigma_{c1}/\sigma_{c2}$. It is assumed that compressive stresses are negative and tensile stresses are positive, and the principal directions are chosen so that $\sigma_{c1} \geq \sigma_{c2}$ algebraically.

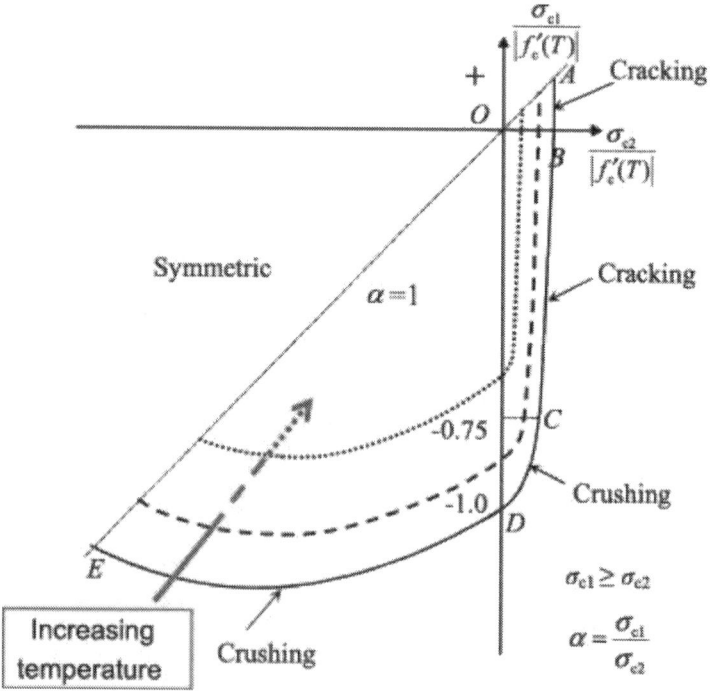

Figure 5. Concrete biaxial failure envelopes at elevated temperatures.

The four regions of the failure surfaces of the strength envelope in Fig. 5 can be expressed as follows:

(1). In the tension–tension region, $(\sigma_{c1} = tension,\ \sigma_{c2} = tension)$, line segment A–B, failure by cracking:

$$\left.\begin{array}{l} \sigma_{c1} = f_t \\ \sigma_{c2} = \frac{\sigma_{c1}}{\alpha} \end{array}\right\} \quad \alpha \geqslant 1.0$$

(2). In the tension–compression region $(\sigma_{c1} = tension, \sigma_{c2} = compression)$, line segment B–C, failure by cracking:

$$\left.\begin{array}{l} \sigma_{c1} = \alpha\sigma_{c2} \\ \sigma_{c2} = \frac{f_t'}{\alpha - 0.6r} \end{array}\right\} \quad \alpha \leqslant -0.73r$$

where $r = f_t' / |f_c'|$.

(3). In the tension–compression region $(\sigma_{c1} = tension, \sigma_{c2} = compression)$, line segment C–D, failure by crushing:

$$\left.\begin{array}{l} \sigma_{c1} = \alpha\sigma_{c2} \\ \sigma_{c2} = \frac{f_c'}{12.8r}\left[9r + \alpha + \sqrt{(9r + \alpha)^2 - 66.56r^2}\right] \end{array}\right\} \quad -0.73r < \alpha \leqslant 0$$

(4). In the compression–compression region $(\sigma_{c1} = compression, \sigma_{c2} = compression)$, line segment D–E, failure by crushing:

$$\left.\begin{array}{l} \sigma_{c1} = \alpha\sigma_{c2} \\ \sigma_{c2} = \frac{1 + 3.65\alpha}{(1+\alpha)^2}f_c' \end{array}\right\} \quad 0 < \alpha \leqslant 1$$

Within this model the initiation of a cracking or crushing process at any location occurs when the concrete stresses reach one of the failure surfaces. It is also assumed that after concrete crushing, all strength and stiffness are lost. The main advantages of this model are that it is simple and the required data are readily obtainable from uniaxial tests on the concrete.

As shown in Fig. 6, the models specified in EN1992-1-2 [48] are adopted to determinate the uniaxial properties of concrete at elevated temperatures. The uniaxial tensile strength of concrete (in MPa) is obtained by $f_t'(T) = 0.3321\sqrt{f_c'(T)}$[49]. Therefore, the concrete tensile strength $f_t'(T)$ changes with temperature as well. The thermal elongation of concrete is calculated according to the model suggested by EN1992-1-2 [48]. For concrete in the biaxial stresses case, it is assumed that free thermal expansion produces zero shear strain.

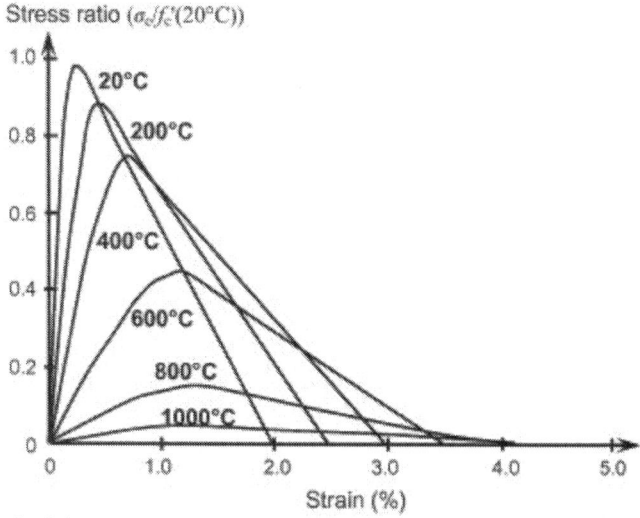

Figure 6. Uniaxial compressive stress–strain relationships of concrete at elevated temperatures.

The Determination of Enhancement Elements and Nodes

Under fire conditions, each concrete layer within an element has different temperatures and material properties. The magnitude and orientation of principle stresses at a Gauss point may also not be the same for each layer. Therefore, the failure envelope of concrete at a Gauss point, which is temperature-dependent, may change over the different concrete layers. Thus, a criterion is needed for determining whether or not an element should be enhanced. In this study, the weighted average values of maximum principal stresses and concrete material properties over the element are proposed to examine the initiation of cracks in an element. For an element the weighted average stress in the x direction ($\sigma_{x,ave}$) and the weighted average tensile strength of concrete $f_{t,ave}(T)$ can be expressed as (see Fig. 2):

$$\sigma_{x,ave} = \frac{\sum_{l=1}^{n} \sum_{g=1}^{m} \sigma_{x,g}^{l}(z_{l+1} - z_{l})}{m(z_{n+1} - z_{1})} \tag{26}$$

$$f_{t,ave}(T) = \frac{\sum_{l=1}^{n} f_{t}^{l}(T)(z_{l+1} - z_{l})}{z_{n+1} - z_{1}} \tag{27}$$

where

$\sigma_{x,ave}$ is the average stress in the x direction,

$f_{t,ave}$ (T) is the average tensile strength of concrete,

$\sigma_{x,g}^l$ is the stress in the x direction at the g-th Gauss point of the l-th layer,

$f_t^l(T)$ is the tensile strength of the l-th layer concrete,

m is the total number of Gauss points in each layer,

n is the total layer number of an element, and

$(z_{n+1} - z_1)$ is the total thickness of an element.

Using the same procedure, the weighted average stress in the y direction $(\sigma_{y,}ave)$ and the weighted average shear stress $(\sigma_{xy,} ave)$ can also be calculated. The weighted average principle stresses $\sigma_{p1,}ave$ and $\sigma_{p2,}ave$ can be obtained from $\sigma_{x,}ave, \sigma_{y,}ave$ and $\sigma_{xy,}ave$. Again, the same method is used to calculate the weighted average compressive strength $f_{c,ave}(T)$, and weighted average modulus of elasticity $E_{c,}$ ave (T), for the concrete element. Based on those parameters the biaxial concrete failure envelope (see Fig. 5) can be constructed for each concrete element at each time or temperature step.

At each time or temperature increment, all concrete elements are examined one by one. Once the average principal tensile stresses of a concrete element reach one of the 'average failure surfaces', either in the biaxial tension region or in the combined tension–compression region, a straight crack is inserted through the entire element, and the orientation of the crack is normal to the average maximum tensile principal stress. The initial crack is assumed to go through the centroid point of a quadrilateral element. Then, when the average principal stresses of the next element reach one of the tension failure surfaces, the crack will propagate from the tip of the existing crack into the next element, following the orientation normal to the corresponding average maximum tensile principal stress of the element. Fig. 7 illustrates how a crack initiates and propagates. As can be seen, there are two different possible ways in which an initial crack cuts a quadrilateral element: initial crack 1 in Fig. 7(a) and initial crack 2 in Fig. 7(b), each of which has possibly three crack propagation paths within the next element when the initial crack extends from element 1 into element 2.

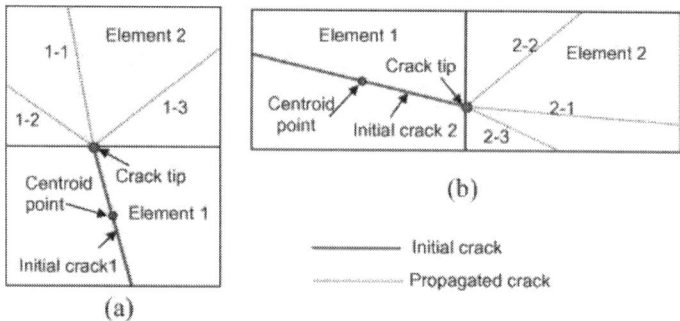

Figure 7. Crack initiation and propagation.

Since the enhancement function (*sign* function) related to enhancement nodes is shifted by $sign\ (x_i)$, the enhanced displacement field vanishes outside the element enclosing the crack. Thus, only the elements crossed by the crack need to be enhanced, rather than all of the elements that contain enhanced nodes. This procedure is illustrated in Fig. 8, where the enhanced elements are filled with grey colour, and the enhanced nodes are indicated by the solid circles and regular nodes by the hollow circles. To model multiple cracks within a reinforced concrete member, the model developed in this paper allows two or more cracks to initiate and propagate at the same time. For simplicity, it is assumed that only one crack may exist within a particular element.

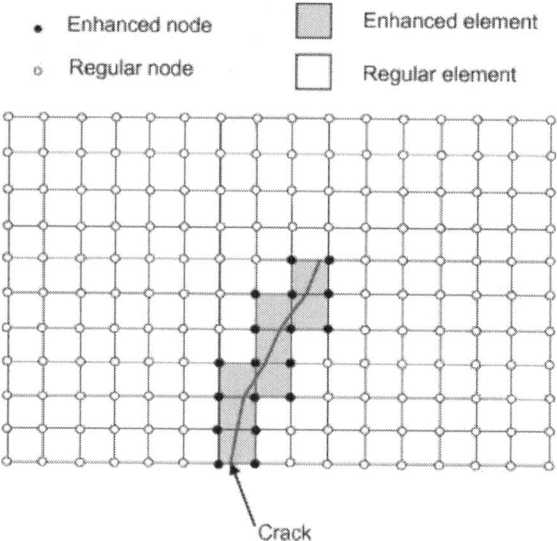

Figure 8. The finite element mesh for a plain concrete structure with a crossed crack.

After concrete cracking, the constitutive model based on the cohesive crack concept [50]is adopted for the cracked concrete element. The crack opening is related to the traction forces acting on the crack. The constitutive model is formed in an orthogonal local coordinate system (n, t) related to the crack, in which the n direction is normal to the crack and the t direction is tangential to the crack. Note that in the cohesive cracks, although the crack faces are not in contact the frictional forces may still exist between the faces due to the applied cohesive stresses if there is relative sliding [51]. However, for the reinforced concrete beam modelled in this paper, the friction between crack faces will not play an important role in the structural behaviour of the beam, due to the fact that the beam is mainly dominated by bending rather than shear. Therefore, the friction between cohesive crack faces was ignored in the current model. The traction–separation law can be expressed as the following equation, in which the crack opening can be obtained from the enhancement nodal displacements related to the enhanced degrees of freedom:

$$\mathbf{t}_{dl} = \mathbf{T}_{dl}\mathbf{w} = \mathbf{T}_{dl}\overline{\mathbf{N}}\ \mathbf{a}_i$$

(28)

where $\mathbf{t}dl$ is the traction in the l -th layer, $\mathbf{T}dl$ is the tangent stiffness matrix of the traction–separation law for the l-th layer, $\mathbf{w}$ is the largest crack opening reached during the loading history, and $\mathbf{a}_i$ is the enhancement nodal displacement. For a cracked element, linear elastic material properties are still assumed in the continuous solid, but the enhancement internal force related to traction over the crack would decrease with the increase of the crack opening. In the cohesive interface the softening curve is governed by the fracture energy (G_f). A concrete bi-linear softening curve is used herein to describe the decrease of traction with the increase of the crack opening after cracking, as shown in Fig. 9, where f'_t is the tensile strength of concrete, tdn is the traction normal to the crack, and w is the crack opening. As tdn is not less than $0.2f'_t$ the relation between traction and opening is given as $t_{dn} = f'_t - 1.25\frac{f_t^2}{G_f}w$, and after tdn drops below $0.2f'_t$ the relation is given as $t_{dn} = 0.2f'_t - \frac{0.2}{6.16}\frac{f_t^2}{G_f}(w - 0.64w_{ch})$ until the opening attains $0.68wch\left(w_{ch} = \frac{G_f}{f'_t}\right)$. When the crack opening exceeds the traction-free open width $(0.68wch)$, the tangent stiffness is set to zero.

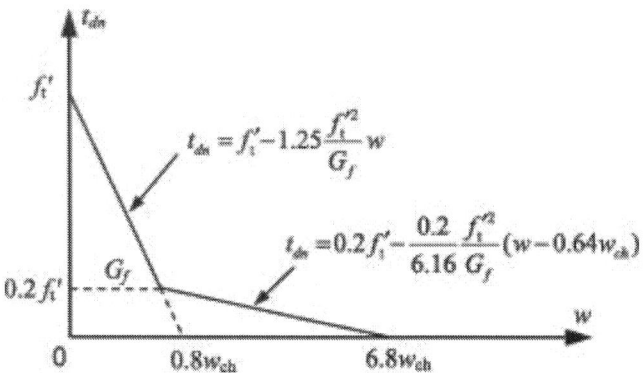

Figure 9. Bi-lineal softening fracture curve of concrete.

Bazant and Becq-Giraudon [52] proposed the fracture energy at an ambient temperature as:

$$G_f(20\ ^\circ C) = 2.5 \times \alpha_0 \left(\frac{f'_c(20\ ^\circ C)}{0.051}\right)^{0.46} \left(1 + \frac{d_a}{11.27}\right)^{0.22} (Ratio_{w/c})^{-0.3} \tag{29}$$

where $\alpha_0 = 1.0$ for rounded aggregates and $\alpha_0 = 1.44$ for crushed or angular aggregates, d_a (in mm) is the maximum aggregate size, and $Ratio_{w/c}$ is the water-to-cement ratio. However, under fire conditions the fracture energy of concrete is expected to change at elevated temperatures. This should be taken into account in the current model. In order to extend the above model to the temperature-dependent cohesive model, the tensile strength (f'_t) and fracture energy (G_f) in Fig. 9 should be temperature-related. The temperature-dependent tensile strength $f'_t(T)$ is calculated using the model specified in EN1992-1-2 [48], and the temperature-dependent fracture energy $G_f(T)$ is determined according to the CEB-FIP model code [53] as:

$$G_f(T) = G_f(20\ ^\circ C)(1.06 - 0.003T) \tag{30}$$

where T is the temperature in $^\circ C$. To replace G_f and f'_t with $G_f(T)$ and $f'_t(T)$ in Fig. 9 respectively, the temperature-dependent cohesive curve is obtained in the current model.

Reinforcing Steel bar and Bond-link Elements

As shown in Fig. 1, a reinforced concrete beam is modelled as an assembly of plain concrete, reinforcing steel bar, and bond-link elements. Previously, a general 3D three-node beam column element was developed by the second author [54], which proved being able to model the reinforced concrete beams and reinforcement bars well. Note that both the 3-node beam element and 2-node element are compatible with the 4-node quadrilateral concrete element in modelling the reinforced concrete structures. For convenience purposes, the three-node beam element developed in [54] is employed in this paper to model the reinforcing steel bar. Each node of the steel bar element contains the conventional six degrees of freedom (three translational and three rotational, in both local and global coordinates). As shown in Fig. 10, the steel mechanical properties and thermal elongation are calculated based on the models specified in EN1992-1-2 [48]. The beam column element allows the reference axis to be placed outside the cross-section. Thus, the reinforcing steel bars can be easily modelled using beam column elements, together with layered concrete elements, to simulate a reinforced concrete beam in fire.

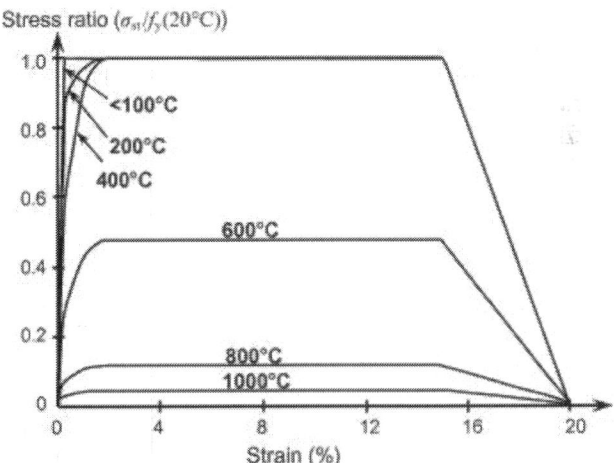

Figure 10. Stress–strain relationships of reinforcing steel at elevated temperatures.

At present there are several ways in which to model the concrete–reinforcement interface, such as the bond-link element and cohesive crack model using the enrichment function. However, the bond-link element has been widely used for structural analysis and design of reinforced concrete structures. Therefore, in order to model the bond characteristic between the concrete and reinforcing steel bar in fire, a two-node bond-link element developed by Huang [55] is employed in this research to link the nodes between a plain concrete element and reinforcing steel bar element. The bond-link element has no physical dimensions, and the two connected nodes originally occupy the same location in the finite element mesh of the undeformed structure. Three bond-link elements are used to connect two plain concrete elements with one 3-node steel bar element. Each node of the bond-link element includes three translational and three rotational degrees of freedom. It is, however, assumed that the slip between reinforcing steel and concrete is related only to the longitudinal axis direction of the steel bar element. The bond element is capable of modelling full, partial and zero bonds between the concrete and reinforcing steel within the reinforced concrete structures.

In order to investigate the mesh sensitivity of the current model a simply supported reinforced concrete beam at an ambient temperature was modelled using different meshes, i.e. 200 concrete elements plus 40 steel bar elements and 400 concrete elements plus 80 steel bar elements. The beam with a length of 2000 mm and a cross-sectional dimension of 150 mm in width and 200 mm in height is reinforced by two ribbed 16 mm (in diameter) tensile steel bars and two ribbed 10 mm (in diameter) compressive steel bars. The compressive strength of concrete at testing is 23.8 MPa. The yield strengths of the 16 mm (in diameter) bar and 10 mm (in diameter) bar are 406 MPa and 365 MPa, respectively. The beam was modelled under four-point loads. The comparison of predicted loads versus mid-span deflection by using different FE meshes is given inFig. 11. It can be seen that the results are almost identical to each other. Therefore, the current model is not very mesh-sensitive.

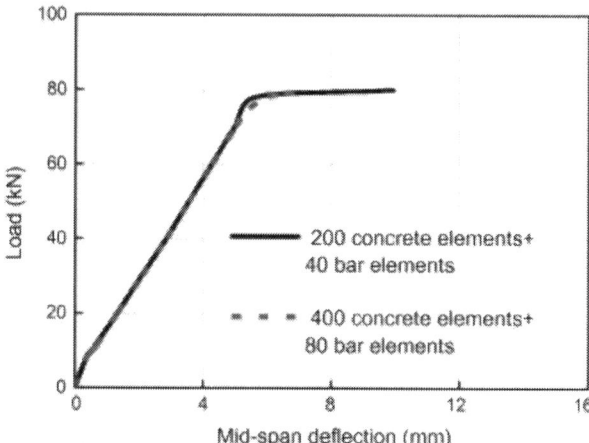

Figure 11. Comparison of predicted loads versus mid-span deflection curve using different FE meshes.

NUMERICAL EXAMPLE AND VALIDATIONS

It is noted that under fire conditions the temperature distribution within the reinforced concrete beam may be significantly affected due to the formation of big localised cracks. In particular, some major cracks may result in the main reinforcing steel bars being directly exposed to fire. However, it is a difficult task to precisely predict the impact of localised cracks on the thermal behaviour of the beam in fire, and this is outside the scope of the current paper. According to the experimental investigations conducted recently by Ervine [56] and [57], in which the rate of thermal propagation through the undamaged beams was compared with the beams with minor cracking (surface crack opening was around 1 mm) and the beams with major cracking (surface crack opening up to around 5 mm), the effect of tensile cracking on the thermal propagation of the beam was not significant and could be ignored in structural analyses. Thus, in this paper, for simplicity, the impact of localised cracks on the thermal behaviour of the beam is not taken into account. However, previous research has indicated that the thermo-hygro-mechanical effect of concrete at a high temperature is significant, especially when concrete is spalled under fire conditions [58], [59], [60], [61] and [62]. The second author of this paper successfully modelled the effects of concrete spalling on the thermal and structural behaviour of reinforced concrete slabs by using a layer procedure to allow some concrete layers to be "void" (with zero mechanical strength and stiffness; zero thermal resistance) [63]. This method can be easily incorporated into the current layered quadrilateral concrete element procedure to model the impact of concrete spalling. For modelling a reinforced concrete beam in fire, the first step of the analysis is to perform the thermal analyses on the beams modelled. Huang et al. [64] developed a two-dimensional nonlinear finite-element procedure (FPRCBC-T) to predict the temperature distributions within the cross-sections of reinforced concrete members subjected to a given fire time–temperature curve. In this study the program FPRCBC-T is used to obtain the temperature history across the section of reinforced concrete beams. The influence of moisture on the concrete is considered. However, the influence of concrete cracking on the temperature distribution is not included. The predicted temperature histories are then used to perform structural analysis for the reinforced concrete beams. The mesh of the cross-sections of the beams used for thermal analysis is also used for the structural analysis. It is assumed that changes in loads or temperatures occur only at the beginning of each time or temperature step. During each step the external loads and temperatures in the layers of all elements are assumed to remain constant.

A Simply Supported Reinforced Concrete Beam in Fire

As a numerical example, a simply supported reinforced concrete beam (subjected to ISO834 standard fire) was modelled to demonstrate the capability of the current model developed for capturing the localised fracture of reinforced concrete beams in fire. Fig. 12 shows the details of the modelled beam. The beam was reinforced by two ribbed 16 mm (in diameter) tensile steel bars and two ribbed 10 mm (in diameter) compressive steel bars. The compressive strength of concrete was 23.8 MPa. The yield strengths of the 16 mm (in diameter) bar and 10 mm (in diameter) bar were 406 MPa and 365 MPa respectively. The transverse point load at mid-span was 40 kN, which was kept constant during the fire. Three cases of a perfect bond, partial bond with ribbed steel bars, and partial bond with smooth steel bars were modelled.

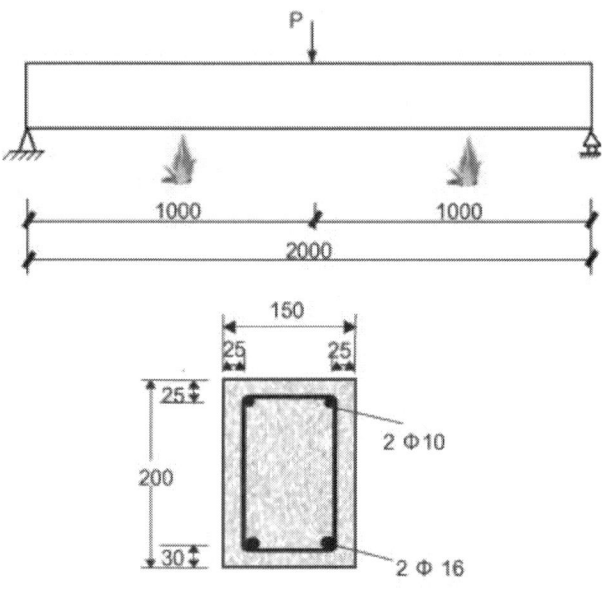

Figure 12. Details of a simply supported reinforced concrete beam under ISO834 fire.

Fig. 13 presents the predicted mid-span deflections of the beam against time for different bond characteristics. For the perfect bond case the deflection–time relation can be generally characterised as four stages before the beam reaches failure, as shown in Fig. 13. At stage 1, the deflection of the beam developed slowly, mainly due to the thermal bowing. The degradation of materials caused by rising temperatures was not remarkable in terms of strength and Young's modulus during this period. At stage 2, the deflection–time curve shows a flat segment from

13 min to 17 min. This is because the temperature of tensile reinforcing steel bars remained almost constant, due to the effect of free water evaporation within the concrete. At stage 3, the temperature of tensile steel bars rose from 120 °C to more than 500 °C, and the deflection rate was increased due to the remarkable deterioration of material strengths at high temperatures. After the tensile reinforcing steel bars yielded (at stage 4), the deflection of the beam increased significantly until the fracture of the steel bars at mid-span. It is obvious that the bond-slip characteristic, between the reinforcing steel bar and the concrete, has a significant influence on the behaviour of reinforced concrete beams in terms of fire resistance and deflection. For instance, the fire resistances of the perfect bond and ribbed bar cases were around 66 min and 62 min, respectively. However, the fire resistance of the smooth bar case was only 13 min.

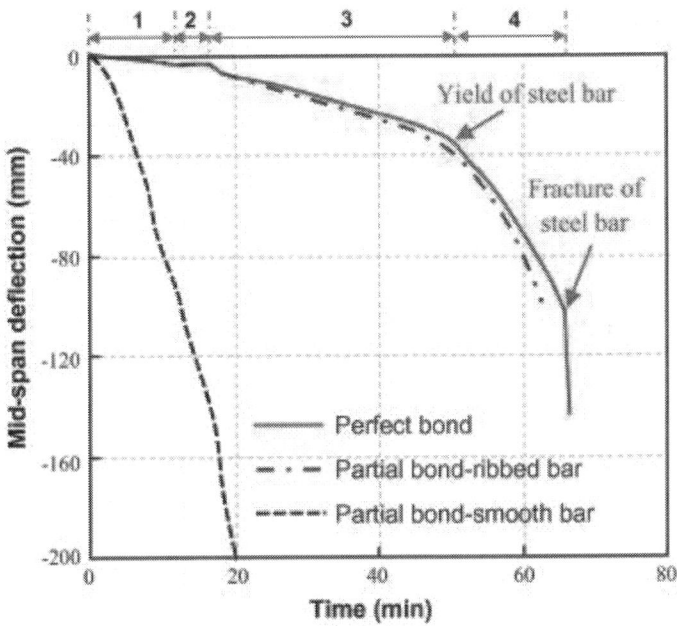

Figure 13. Predicted mid-span deflections with time for different bond characteristics.

Fig. 14 shows the predicted crack opening history at mid-span of the beam, in which the crack openings were calculated using Eq. (28). It was found that the crack opening at mid-span decreased slightly at the initial stage due to the effect of thermal expansion, and then the crack opening increased gradually. After the steel bars yielded, the crack opening increased significantly until the fracture of the steel bars. It is obvious that the bond characteristic between the reinforcing steel bar and concrete has a significant influence on the maximum crack opening. Under the same fire exposure time, the opening of the ribbed bar case was greater than that of the perfect bond case, and the difference between the two cases became more significant after the tensile steel bars yielded. In the smooth bar case, its maximum crack opening had already exceeded 15 mm at the ambient temperature before the fire, and the crack opening of the smooth bar case increased dramatically at elevated temperatures.

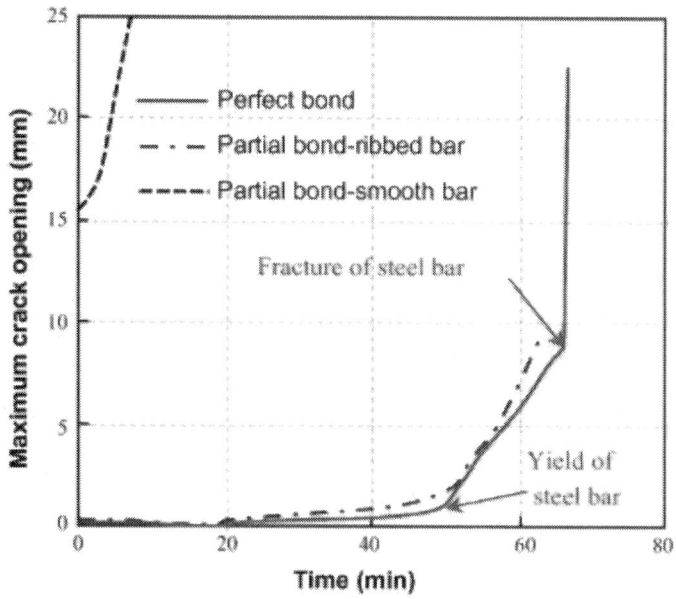

Figure 14. Predicted maximum crack openings with time for different bond characteristics.

The cracking pattern and deformed mesh of the beam, with a perfect bond condition at 66 min of fire time, are shown in Fig. 15(a) and (b). It can be seen that the XFEM developed in this paper can reasonably predict the formation and propagation of individual cracks. For the perfect bond case at 66 min of fire time, the openings of two major cracks at mid-span were 23.0 mm and 17.4 mm respectively. However, the cracks near the supports were only 0.14 mm and 0.16 mm, respectively. As shown in Fig. 15(c), for the smooth bar case, the maximum opening of the major crack at mid-span reached 94.6 mm. In both cases the mid-span elements appear obviously distorted, because the mesh held a very big localised crack. It is evident that the model proposed is able to capture the localised fracture of reinforced concrete beams under fire conditions very well.

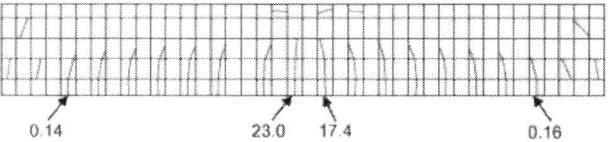

(a) Predicted cracking pattern (perfect bond)

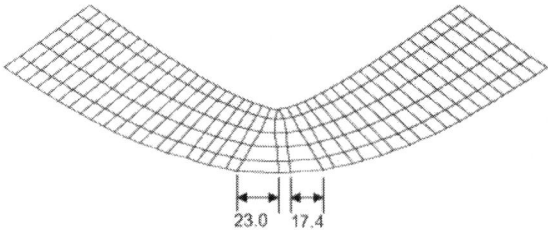

(b) Predicted deformed mesh (perfect bond, x-axis displacement has been amplified 5 times).

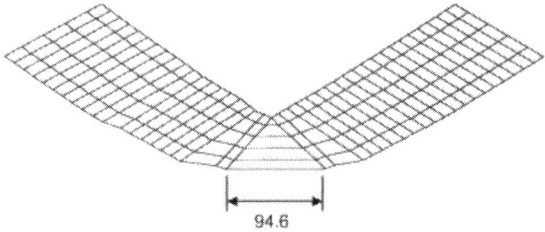

(c) Predicted deformed mesh (partial bond with smooth bars, x-axis displacement has been amplified 5 times).

Figure 15. Predicted localised cracks of reinforced concrete beams in fire for different bond conditions (unit: mm).

Fire Tests of Reinforced Concrete Beams

In order to validate the model proposed in this paper, four reinforced concrete beams subjected to fire tests were modelled herein. These fire tests on normal-strength reinforced concrete beams with ribbed steel bars were conducted by Lin et al. [65]. For these tests, two heating curves, the ASTM fire and Short Duration High Intensity (SDHI) fire, were adopted. Here the four beams, designated as Beams 1, 3, 5 and 6, were modelled. Beams 1 and 3 were heated using the ASTM fire, and Beams 5 and 6 were subjected to the SDHI fire. Fig. 16 provides details of Beams 1, 3, 5 and 6, where the load$P$ was kept constant at 44.48 kN during each fire test, although the cantilever force P_o varied as the test progressed. The history of the cantilever force P_o is illustrated in Fig. 17. The measured material properties at room temperature were the: concrete's compressive strengths: $f'_c(20\,^\circ\text{C}) = 27.86$ MPa (Beam 1), $f'_c(20\,^\circ\text{C}) = 31.5$ MPa (Beam 3), $f'_c(20\,^\circ\text{C}) = 33.73$ MPa (Beam 5), $f'_c(20\,^\circ\text{C}) = 34.54$ MPa (Beam 6), and steel yield strengths: $f_t(20\,^\circ\text{C}) = 487.27$ MPa (bar #7), and $f_t(20\,^\circ\text{C}) = 509.54$ MPa (bar #8). Those tested material properties were used for the validations.

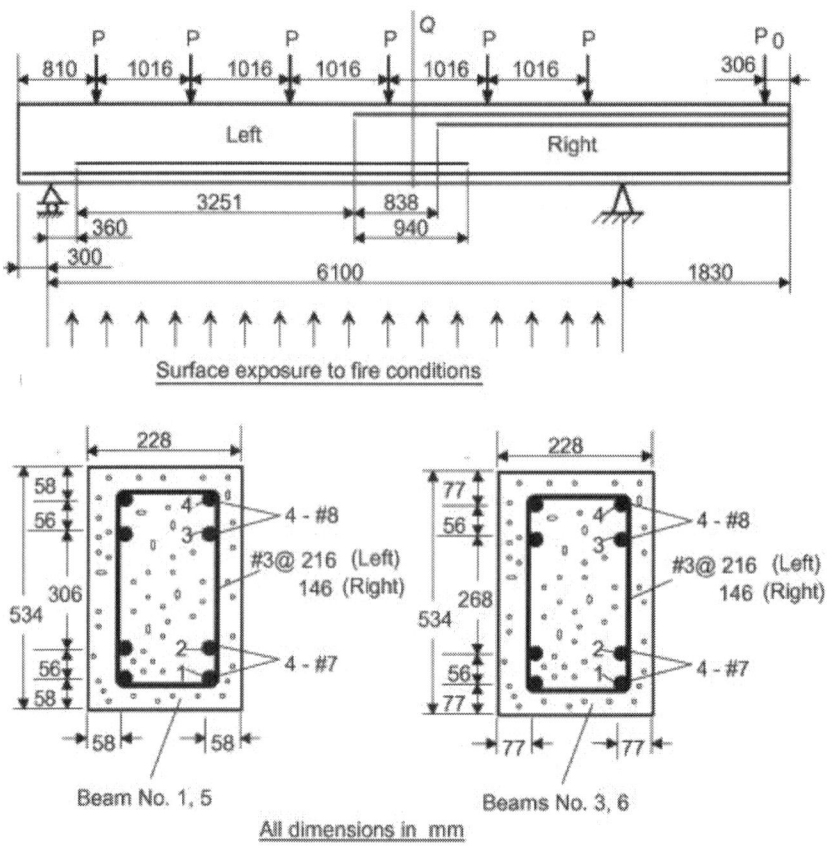

Figure 16. Details of tested reinforced concrete beams in fires (adapted from [65]).

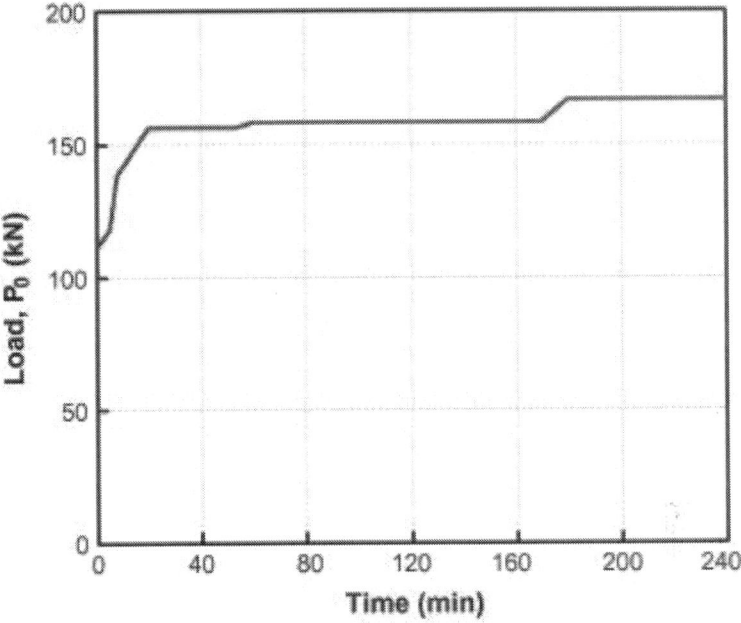

Figure 17. History of load (P_0) at cantilever end (adapted from [65]).

The tested beams were subjected to three-face heating from its bottom and two sides. For the thermal analysis the cross-section of beams is divided, as 12 columns by 14 rows, with a total of 168 segments. This means that each concrete element was divided into 12 layers for the structural analysis. As an example, Fig. 18 shows the division of the cross-section of Beam 1, where the size of the concrete segments close to the fire boundary is less than that of the concrete segments away from the fire. Fig. 19 illustrates the predicted temperatures, together with tested results for four main reinforcing steel bar layers of Beams 1 and 5, where the reinforcing steel layers are denoted in sequence from bottom to top as Layers 1–4 (see Fig. 16). It is evident that good agreement has been achieved between the tests and predictions. The temperature histories obtained in the thermal analysis were subsequently used for the structural analyses of the beams.

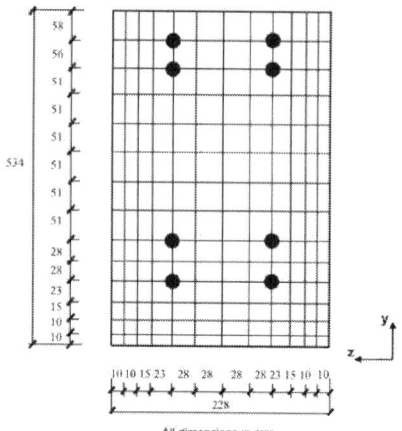

Figure 18. The division of the cross-section of Beam 1 for the thermal analysis.

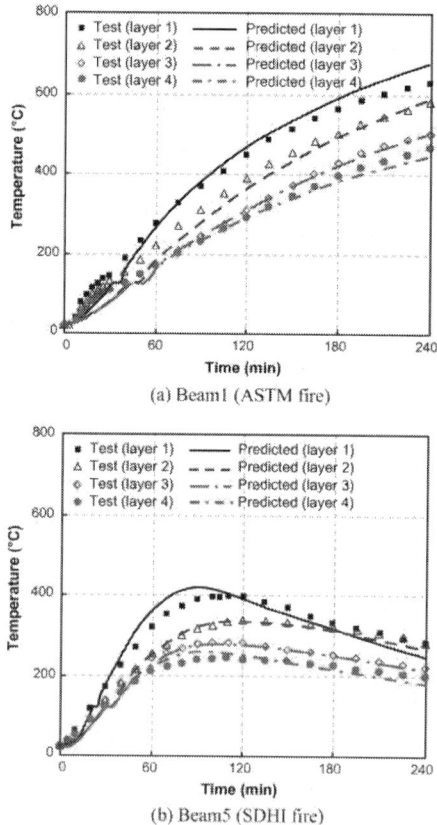

(a) Beam1 (ASTM fire)

(b) Beam5 (SDHI fire)

Figure 19. Comparison of predicted and measured temperatures of four main reinforcing steel bars of Beams 1 and 5.

In the analysis, each beam was modelled as an assembly of 1652 (14 × 118) layered quadrilateral plain concrete elements, 146 steel bar beam elements, and 296 bond-link elements. The mesh sensitivity test was conducted before the analysis, where the doubly finer mesh was used for the comparison. The results of the current mesh and the finer mesh were almost identical to each other. It is evident that the predicted results are not sensitive to the element size under the current mesh used. The predicted maximum vertical deflections for Beams 1, 3, 5 and 6, against time, are presented in Fig. 20, Fig. 21, Fig. 22 and Fig. 23. The maximum vertical deflections of all beams appeared at a position about 3500 mm away from the right-hand side support (see Fig. 16). It is evident that the predictions of the developed model agree reasonably well with the test results for the four beams in terms of deflections. It was found that the beams showed small upward deflections at the initial stage of the fire. This was due to the fact that the cantilever force P_o kept increasing at the initial stage (see Fig. 17), which tended to result in the downward deflection at the cantilever end, and the corresponding upward deflection at the first bay of beams. Afterwards, when P_o was kept stable, the beams developed downward deflections due to the influence of elevated temperatures.

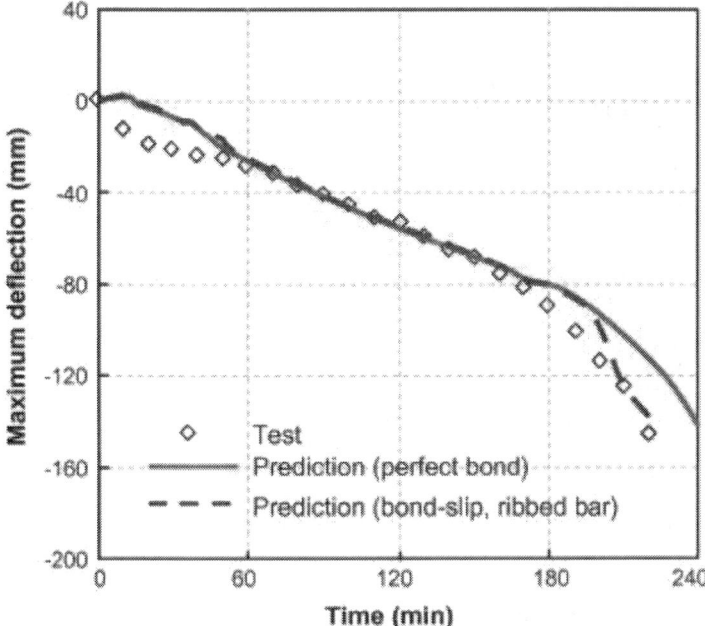

Figure 20. Comparison of predicted and measured maximum deflections of Beam 1 (ASTM fire).

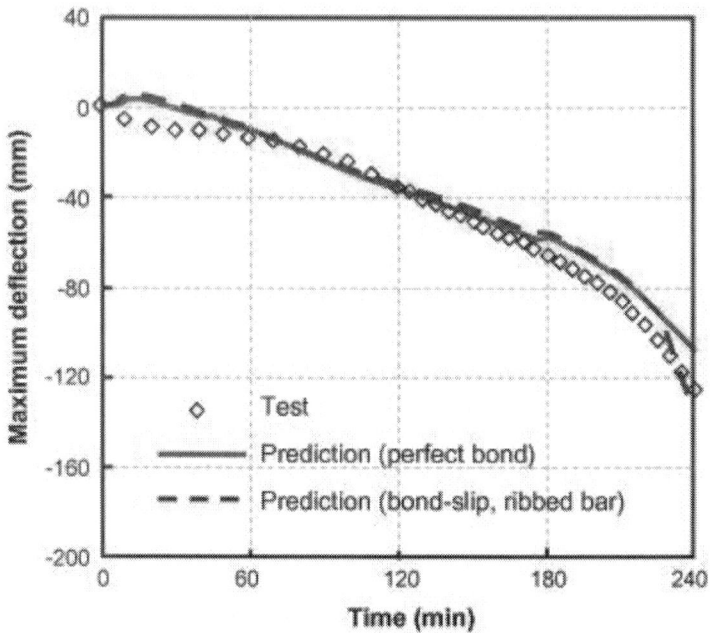

Figure 21. Comparison of predicted and measured maximum deflections of Beam 3 (ASTM fire).

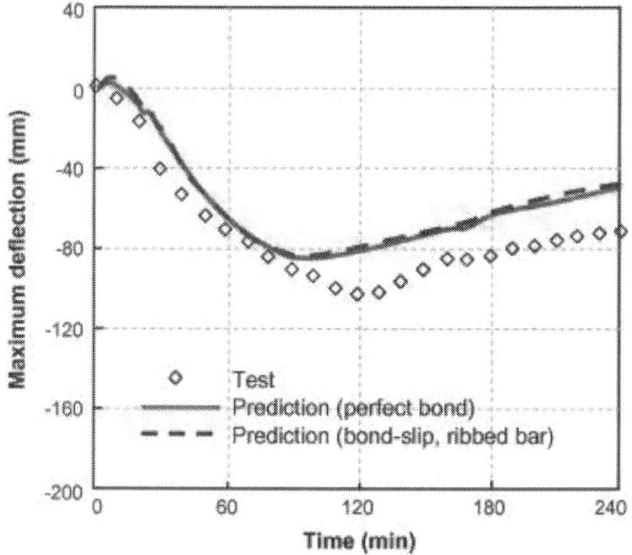

Figure 22. Comparison of predicted and measured maximum deflections of Beam 5 (SDHI fire).

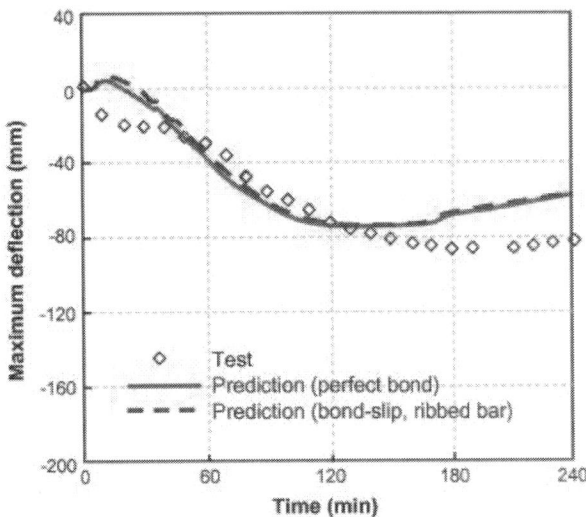

Figure 23. Comparison of predicted and measured maximum deflections of Beam 6 (SDHI fire).

From Fig. 20 and Fig. 21 it can be seen that for Beams 1 and 3 there is no obvious difference between the ribbed bar case and perfect bond case in terms of deflections until the later stage of fire. Two predicted deflection curves diverted: after 180 min for Beam 1, and 225 min for Beam 3. It is evident that the bond characteristics have a considerable influence on these two beams in terms of the deflections at the later stages of a fire. Generally, compared to the perfect bond case, the predicted deflections by the ribbed bar case agreed better with the test results. Therefore, if a perfect bond condition is assumed for modelling the interaction between reinforcing steel bars and concrete, the predicted results may be on the unconservative side.

The results of Beams 5 and 6 are respectively shown in Fig. 22 and Fig. 23. It can be seen that the effect of bond characteristics on the deflection of the mid-span areas of the beams is relatively small, compared to Beams 1 and 3. This is due to the fact that these beams were subjected to a Short Duration High Intensity (SDHI) fire, and the maximum temperatures of the reinforcing steel bar did not exceed 400 °C (see Fig. 19). It is also interesting to see that the deflection of the ribbed bar case is even slightly smaller than that of the perfect bond case. This is owing to the fact that the ribbed bar case has slightly bigger upward deflections than the perfect bond case at the initial stages of the fire. However, from the deflection of the cantilever end (shown in Fig. 24) for Beam 6, it can be seen that bond conditions have a more significant influence on the continuous support than the mid-span areas of the beam.

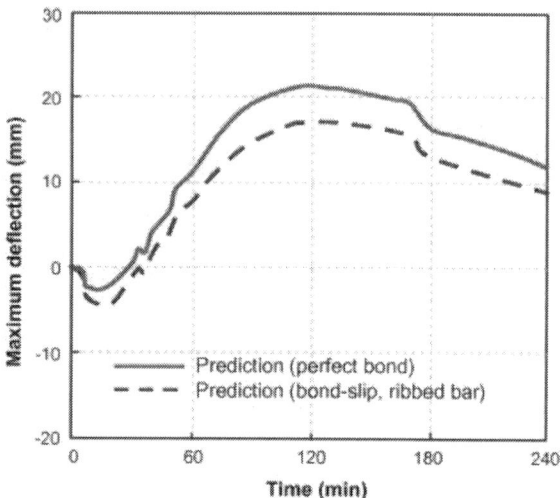

Figure 24. Comparison of predicted deflections at the cantilever end of Beam 6 (SDHI fire).

Fig. 25(a) and (b) present the predicted cracking patterns of Beams 1 and 5, respectively. It is evident that the current model can predict the formation and propagation of individual cracks quite well. The predicted crack patterns are reasonable where the flexural cracks caused by a sagging moment distribute at the lower part of mid-span areas, whilst the flexural cracks caused by a hogging moment distribute at the upper part over the continuous support. Besides that, diagonal cracks were also found within the reinforced concrete beam. The maximum crack opening of Beam 1 attained 6.05 mm near the mid-span, and the maximum crack opening of Beam 5 reached 0.65 mm only due to the Short Duration High Intensity (SDHI) fire applied.

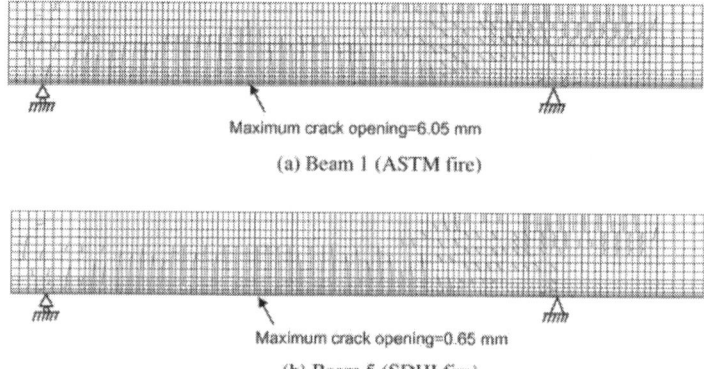

Maximum crack opening=6.05 mm

(a) Beam 1 (ASTM fire)

Maximum crack opening=0.65 mm

(b) Beam 5 (SDHI fire)

Figure 25. Predicted cracking patterns of Beams 1 and 5.

CONCLUSIONS

In this paper a robust layered finite element procedure is proposed for modelling the localised fracture of reinforced concrete beams at elevated temperatures. In this new model the plain concrete is modelled by 4-node layered quadrilateral elements, incorporated into the extended finite element method (XFEM). The element is divided into layers to take into account the temperature distribution over the cross-section of the beam. Additional degrees of freedom are used to describe a discontinuous displacement field, and the enhancement function is used to realise the displacement jump over a crack for the cracked elements. A criterion based on a weighted average stress approach is proposed to determine the initiation of individual cracks within the plain concrete elements. The complications of structural behaviour under fire conditions, such as thermal expansion, the bond characteristic between reinforcing steel bars, concrete at elevated temperatures, and the change of material properties with temperature, are all considered in the model.

The new model has been validated against previous fire test results on reinforced concrete beams. A numerical example of modelling a simply supported reinforced concrete beam (subjected to ISO834 fire) has been analysed to demonstrate the capability of the current model for capturing the localised fracture of reinforced concrete beams under fire conditions. It has been shown that the XFEM nonlinear procedure proposed can predict the global response of reinforced concrete beams with good accuracy. The formation and propagation of individual cracks within the beams are also modelled, capturing the localised fracture, and predicting crack openings during the analysis. The model developed in this paper provides an excellent numerical approach for assessing both structural stability (global behaviour) and integrity (localised fracture) of reinforced concrete members in fire. The model proposed here will be further extended to 3D modelling of localised fracture of reinforced concrete slabs under fire conditions in order to assess the integrity failure of concrete floor slabs.

ACKNOWLEDGEMENT

The authors gratefully acknowledge the support of the Engineering and Physical Sciences Research Council of Great Britain under Grant No. EP/I031553/1.

REFERENCES

1. Bailey CG, Toh WS. Small-scale concrete slab tests at ambient and elevated temperatures. Eng Struct 2007;29(10):2775–91.
2. Foster SJ. Tensile membrane action of reinforced concrete slabs at ambient and elevated temperatures. Ph.D. Thesis. Department of Civil and Structural Engineering, University of Sheffield; 2006.
3. Foster SJ, Bailey CG, Burgess IW, Plank RJ. Experimental behaviour of concrete floor slabs at large displacement. Eng Struct 2004;26(9):1231–47.
4. Wang Y, Dong YL, Zhou GC. Nonlinear numerical modeling of two-way reinforced concrete slabs subjected to fire. Comput Struct 2013;119(4):23–36.
5. Davie CT, Zhang HL, Gibson A. Investigation of a continuum damage model as an indicator for the prediction of spalling in fire exposed concrete. Comput Struct 2012;94–95(3):54–69.
6. Chung JH, Consolazio GR, McVay MC. Finite element stress analysis of a reinforced high-strength concrete column in severe fires. Comput Struct 2006;84(8):1338–52.
7. Capua DD, Mari AR. Nonlinear analysis of reinforced concrete cross-sections exposed to fire. Fire Saf J 2007;42(2):139–49.
8. Kodur VKR, Dwaikat M. A numerical model for predicting the fire resistance of reinforced concrete beams. Cem Concr Compos 2008;30(5):431–43.
9. Riva P, Franssen JM. Non-linear and plastic analysis of RC beams subjected to fire. Struct Concr 2008;9(1):30–43.
10. Izzuddin BA, Elghazouli AY. Failure of lightly reinforced concrete members under fire. I: Analytical modelling. J Struct Eng ASCE 2004;130(1):3–17.
11. Huang ZH, Burgess IW, Plank RJ. Nonlinear analysis of reinforced concrete slabs subjected to fires. ACI Struct J 1999;96(1):127–35.
12. Huang ZH, Burgess IW, Plank RJ. Modelling membrane action of concrete slabs in composite buildings in fire. Part I: Theoretical development. J Struct Eng ASCE 2003;129(8):1093–102.

13. Bratina S, C̆as B, Saje M, Planinc I. Numerical modelling of behaviour of reinforced concrete columns in fire and comparison with Eurocode 2. Int J Solids Struct 2005;42(10):5715–33.

14. Gao WY, Dai JG, Teng JG, Chen GM. Finite element modeling of reinforced concrete beams exposed to fire. Eng Struct 2013;52(7):488–501.

15. Belytschko T, Black T. Elastic crack growth in finite elements with minimal remeshing. Int J Numer Meth Eng 1999;45(5):601–20.

16. Moës N, Dolbow J, Belytschko T. A finite element method for crack growth without remeshing. Int J Numer Meth Eng 1999;46(1):131–50.

17. Melenk JM, Babus̆ka I. The partition of unity finite element method: basic theory and application. Comput Meth Appl Mech Eng 1996;139(1–4): 289–314.

18. Wells GN, Sluys LJ. A new method for modelling cohesive cracks using finite elements. Int J Numer Meth Eng 2001;50(12):2667–82.

19. Zi G, Belytschko T. New crack-tip elements for XFEM and applications to cohesive cracks. Int J Numer Meth Eng 2003;57(15):2221–40.

20. Verhoosel CV, Remmers JJC, Gutiérrez MA. A dissipation-based arc-length method for robust simulation of brittle and ductile failure. Int J Numer Meth Eng 2009;77(9):1290–321.

21. Moës N, Belytschko T. Extended finite element method for cohesive crack growth. Eng Fract Mech 2002;69(7):813–33.

22. Hansbo A, Hansbo P. A finite element method for the simulation of strong and weak discontinuities in solid mechanics. Comput Methods Appl Mech Eng 2004;193(33–35):3523–40.

23. Budyn É, Zi G, Moës N, Belytschko T. A method for multiple crack growth in brittle materials without remeshing. Int J Numer Meth Eng 2004;61(10): 1741–70.

24. Daux C, Moës N, Dolbow J, Sukumar N, Belytschko T. Arbitrary branched and intersecting cracks with the extended finite element method. Int J Numer Meth Eng 2000;48(12):1741–60.

25. Réthoré J, Gravouil A, Combescure A. An energy-conserving scheme for dynamic crack growth using the extended finite element method. Int J Numer Meth Eng 2005;63(5):631–59.

26. Duan Q, Song JH, Menouillard T, Belytschko T. Element-local level set method for three-dimensional dynamic crack growth. Int J Numer Meth Eng 2009;80(12):1520–43.

27. Bordas S, Nguyen PV, Dunant C, Guidoum A, Nguyen-Dang H. An extended finite element library. Int J Numer Meth Eng 2007;71(6):703–32.

28. Talebi H, Silani M, Bordas S, Kerfriden P, Rabczuk T. A computational library for multiscale modeling of material failure. Comput Mech 2014;53(5):1047–71.

29. Natarajan S, Bordas S, Mahapatra DY. Numerical integration over arbitrary polygonal domains based on Schwarz–Christoffel conformal mapping. Int J Numer Meth Eng 2009;80(1):103–34.

30. Bordas S, Rabczuk T, Hung NX, Nguyen VP, Natarajan S, Bog T, et al. Strain smoothing in FEM and XFEM. Comput Struct 2010;88(23–24):1419–43.

31. Chen L, Rabczuk T, Bordas SP, Liu G, Zeng K, Kerfriden P. Extended finite element method with edge-based strain smoothing (ESm-XFEM) for linear elastic crack growth. Comput Methods Appl Mech Eng 2012;209–212(2):250–65.

32. Vu-Bac N, Nguyen-Xuan H, Chen L, Bordas S, Kerfriden P, Simpson RN, et al. A node-based smoothed extended finite element method (NS-XFEM) for fracture analysis. Comput Model Eng Sci 2011;73(4):331–56.

33. Bordas S, Natarajan S, Kerfriden P, Augarde CE, Mahapatra DR, Rabczuk T, et al. On the performance of strain smoothing for quadratic and enriched finite element approximations (XFEM/GFEM/PUFEM). Int J Numer Meth Eng 2011;86(4–5):637–66.

34. Dias-da-Costa D, Alfaiate J, Sluys LJ, Areias P, Júlio E. An embedded formulation with conforming finite elements to capture strong discontinuities. Int J Numer Meth Eng 2013;93(2):224–44.

35. Simoni L, Schrefler BA. Multi field simulation of fracture. Adv Appl Mech 2014;47:367–519

36. chapter 4 (Editor: S. Bordas)..

37. Rabczuk T, Bordas S, Zi G. On three-dimensional modelling of crack growth using partition of unity methods. Comput Struct 2010;88(23–24):1391–411.
    F. Liao, Z. Huang / Computers and Structures 152 (2015) 11–26 25

38. Nguyen VP, Rabczuk T, Bordas S, Duflot M. Meshless methods: a review and computer implementation aspects. Math Comput Simulat 2008;79(3):763–813.

39. Bordas S, Rabczuk T, Zi G. Three-dimensional crack initiation, propagation, branching and junction in non-linear materials by extrinsic discontinuous enrichment of meshfree methods without asymptotic enrichment. Eng Fract Mech 2008;75(5):943–60.

40. Bordas S, Zi G, Rabczuk T. Three-dimensional non-linear fracture mechanics by enriched meshfree methods without asymptotic enrichment. In: Combescure A, Borst RD, Belytschko T, editors. IUTAM symposium on discretization methods for evolving discontinuities, chapter: meshless finite element methods. Springer; 2007. p. 21–36.

41. Rabczuk T, Zi G, Bordas S, Nguyen-Xuan H. A geometrically non-linear three dimensional cohesive crack method for reinforced concrete structures. Eng Fract Mech 2008;75(16):4740–58.

42. Xiao QZ, Dhanasekar M. Plane hybrid stress elements for 3D analysis of moderately thick solids subjected to loading symmetric to midsurface. Int J Solids Struct 2007;44(7–8):2458–76.

43. Ahmed A. Extended finite element method (XFEM)-modeling arbitrary discontinuities and failure analysis. Master Thesis, Italy: University of Pavia; 2009.

44. Bathe KJ. Finite element procedures. New Jersey: Prentice-Hall Inc.; 1996.

45. Liao FY, Huang ZH. A nonlinear procedure for modelling localized fracture of reinforced concrete structures. Report No. SED-DCE-2003-1, UK: Brunel University; 2013.

46. Xiao QZ, Karihaloo BL. Improving the accuracy of XFEM crack tip fields using higher order quadrature and statically admissible stress recovery. Int J Numer Meth Eng 2006;66(9):1378–410.

47. Barzegar-Jamshidi F. Nonlinear finite element analysis of reinforced concrete under short term monotonic loading. Ph.D. Thesis, University of Illinois at Urbana-Champaign; 1987.

48. Kupfer HB, Gerstle KH. Behavior of concrete under biaxial stresses. J Eng Mech Div ASCE 1973;99(EM4):853–66.

49. EN 1992-1-2. Eurocode 2, design of concrete structures, Part 1.2: General rules – structural fire design. Brussels: Commission of the European Communities; 2004.

50. American Society of Civil Engineers. Finite element analysis of reinforced concrete. New York; 1982.

51. Hillerborg A, Modeer M, Petersson PE. Analysis of crack formation and crack growth in concrete by means of fracture mechanics and finite elements. Cem Concr Res 1976;6(6):773–82.

52. Karihaloo BL, Xiao QZ. Asymptotic fields at the tip of a cohesive crack. Int J Fract 2008;150(1–2):55–74.

53. Bazˇant ZP, Becq-Giraudon E. Statistical prediction of fracture parameters of concrete and implications for choice of testing standard. Cem Concr Res 2002;32(4):529–56.

54. CEB-FIP Model Code: Committee Euro-International du Beton. Bulletin D'information No. 213/214. Tomas Telford (London); May 1993.

55. Huang ZH, Burgess IW, Plank RJ. Three-dimensional analysis of reinforced concrete beam–column structures in fire. J Struct Eng ASCE 2009;135(10): 1201–12.

56. Huang ZH. Modelling the bond between concrete and reinforcing steel in a fire. Eng Struct 2010;32(11):3660–9.

57. Ervine A. Damaged reinforced concrete structures in fire. Ph.D. Thesis. Department of Civil and Structural Engineering, University of Edinburgh; 2012.

58. Ervine A, Gillie M, Stratford TJ, Pankaj P. Thermal propagation through tensile cracks in reinforced concrete. J Mater Civ Eng 2012;24(5):516–22.

59. Baggio P, Majorana CE, Schrefler BA. Thermo-hygro-mechanical analysis of concrete. Int J Numer Meth Fl 1995;20(6):573–95.

60. Bazˇant ZP. Analysis of pore pressure, thermal stress and fracture in rapidly heated concrete. In Phan LT, Carino NJ, Duthinh D, Garboczi E, editors. International workshop on fire performance of high-strength concrete. Gaithersburg (MD); 1997. p. 155–64.

61. Kodur VKR, Dwaikat MB. Hydro thermal model for predicting fire-induced spalling in concrete structural systems. Fire Saf J 2009;44(3):425–34.

62. Mindeguia JC, Pimienta P, Carré H, Borderie CL. Experimental analysis of concrete spalling due to fire exposure. Eur J Environ Civ Eng 2013;17(6): 453–66.

63. Kukla, K. Concrete at high temperatures: hygro-thermo-mechanical degradation of concrete. Ph.D. Thesis. School of Engineering, University of Glasgow; 2010.

64. Huang ZH. The behaviour of reinforced concrete slabs in fire. Fire Saf J 2010;45(5):271–82.

65. Huang ZH, Platten A, Roberts J. Non-linear finite element model to predict temperature histories within reinforced concrete in fires. Build Environ 1996;31(2):109–18.

66. Lin TD, Ellingwood B, Piet O. Flexural and shear behaviour of reinforced concrete beams during fire tests. Report no. NBS-GCR-87-536, Center for Fire Research, National Bureau of Standards; 1987.

CITATION

Feiyu Liao, Zhaohui Huang, An extended finite element model for modelling localised fracture of reinforced concrete beams in fire, Computers & Structures, Volume 152, May 2015, Pages 11-26, ISSN 0045-7949, http://dx.doi.org/10.1016/j.compstruc.2015.02.006.

CHAPTER 7

Finite Element Modeling of Steel Concrete Beam Considering Double Composite Action

Ashraf Mohamed Mahmoud

Department of Civil Engineering, Faculty of Engineering, Modern University for Technology and Information, Cairo, Egypt

ABSTRACT

Steel concrete composite construction has gained wide acceptance as an alternative to pure steel or concrete construction. Ansys 11 computer program has been used to develop a three-dimensional nonlinear finite element model in order to investigate the fracture behaviors of continuous double steel-concrete composite beams, with emphasis on the beam slab interface. Three beam models with varying number of the head studs have been addressed. The associated constitutive results such as the ultimate loads, the maximum deflections, the interface slip and slip strain values are presented. A parametric study has been carried out in order to investigate the effect of some parameters on their fracture capabilities, such as steel beam height, lower slab thickness and length, studs diameter and arrangement method. By comparing these results with the available experimental data, the proposed model is found to be capable of analyzing steel-concrete composite beams to an acceptable accuracy.

INTRODUCTION

The use of composite structures is increasingly present in civil construction works. Steel-concrete composite beams, particularly, are structures consisting of two materials, a steel section located mainly in the tension region and a concrete section, located in the compression cross-sectional area, both connected by metal devices known as shear connectors. One type of these connectors is called head studs as shown in Fig. 1. The main functions of these studs are to allow for the joint behavior of the beam-slab, to restrict longitudinal slipping and uplifting at the elements interface and to take shear forces. Double steel-concrete composite continuous beam is a new structural system developed on the basis of single steel-concrete one, in which there is also a bottom reinforced concrete slab connected to a steel profile in the negative moment regions through the head studs, therefore with two interfaces. Comparing with the traditional single steel-concrete composite continuous beam, its advantage is that effectively limits the crack width of the negative moment area, and also improves the stress state of section, so that it is suitable to the composite continuous beam with a larger span. The mechanical properties of the double composite beam obviously depend on their respective properties and interactions. In the negative applied bending moment area, the concrete slab cracks under tension and then the interface slip occurs between steel profile and concrete slab, with non-linear features, it makes great impact on the structure of the internal forces and deformation. Therefore, it is necessary to present a finite element model to study the mechanical properties of the double steel-concrete composite beam in negative moment regions.

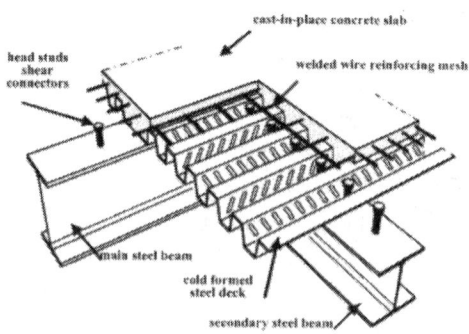

(a) Illustrative sketch of roof slab with composite action.

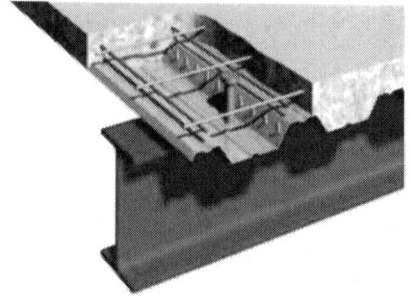

(b) Composite beam system with head studs shear connectors.

Figure 1. Steel-concrete composite section with studs shear connectors.

Although many experimental and theoretical studies for the traditional single steel-concrete composite beam have been done, few research studies have been found in references to the double steel-concrete composite continuous beam. Rozsas [1] investigated the plastic reserve of composite plate girder bridges due to the synergetic combination of the concrete and steel. The plastic design in the framework of the Eurocode through an existing elastically designed bridge is also introduced. Xu et al. [2] discussed the improvement of the local buckling strength of continuous double composite box girders by adding a concrete slab to the steel bottom flange. The mechanical properties in concrete crack, formation of sectional plastic hinge are also investigated. Tan et al. [3] utilized experimental tests to provide further information and conclusions regarding composite steel-concrete beam specimens by examining the behavior of multi-span composite steel-concrete beams. These beams are subjected to combined actions of torsion and flexure for both full and partial shear connection and comparing the disparity in the varying degrees of shear connection. Lin and Yoda [4]studied the mechanical performance of the horizontally curved continuous composite steel-concrete beams subjected to combined hogging (negative) bending and torsion, in order to investigate the effect of curvature on both elastic and inelastic behaviors of these beams in the interior support regions. Henriques et al. [5] presented a generalized beam theory (GBT) formulations specially designed for performing efficient linear analysis of steel-concrete composite bridges and elastoplastic collapse analysis of thin-walled steel members and extended for including the non-linear reinforced concrete material behavior of steel-concrete composite beams. Liang et al. [6] have undertaken nonlinear finite element analysis on continuous composite beams in combined bending and shear. In their study, design formulas incorporating contributions from the concrete slab and composite action were proposed for vertical shear strength and the ultimate shear interaction of continuous composite beams. A finite element model is presented by Liang et al. [7] to investigate the flexural and shear strengths of simply supported composite beams under combined bending and shear. In this research, the numerical results are verified and compared with the available experimental results. Sebastian and McConnel [8] described a nonlinear finite element program for modeling composite beams. Axial springs with empirical shear slip relations were used to model discrete shear connectors. Hirst and Yeo [9] used a standard finite element program to analyze composite beams with partial and full shear connection. Quadrilateral elements were employed to simulate discrete and stud shear connectors. The material properties of stud elements were modified to make them equivalent in strength and stiffness to the actual shear connectors in composite beams. Al-Amery and Roberts [10] presented a nonlinear analysis of composite beams with partial shear connection by using a finite difference method. Salari et al. [11] formulated a composite beam element based on the force analysis method for the nonlinear analysis of composite beams with deformable shear connectors. Thevendran et al. [12] utilized the finite element software ABAQUS to study the ultimate load behavior of composite beams curved in plan. Shell elements were used to model the concrete slab and the steel beam while a rigid beam element was employed to simulate stud shear connectors. Reiner [13] and Stroh and Sen [14] presented a double steel-concrete composite continuous beam as a new structural system developed on the basis of single steel-concrete composite beam, in which there is also a bottom reinforced concrete slab connected to a steel profile in the negative moment regions through the shear connectors, therefore with two interfaces. This research was accompanied by the determination of the crack width limits of the negative moment area, and the improvement of the stress state of section, and later applied for the composite continuous beam with a larger span. Newmark et al. [15] introduced the partial collaboration theory which is used later for deriving the elastic stiffness matrix in the negative moment region for a double composite beam element and for studying and verifying the double composite continuous beam models, and consequently the composite action effect as illustrated by Duan et al. [16], [17] and [18]. Nagai et al. [19] tested a double composite girder under pure hogging moment and measured its ultimate bending moment strength. Duan et al. [20] and Yang and Duan [21] focused on the problems of interface slip, deformation, ultimate bearing capacity, and the effective flange width of concrete slab for the double steel-concrete composite beams. Wang et al. [22] presented the elastic analysis of double composite beam deformations using the Goodman elastic sandwich method. Yen et al. [23] discussed the ultimate load behavior and elastic deformations of steel box girders containing composite bottom flanges. Duan et al. [24] performed beam collapse tests for three models of double steel-concrete composite continuous beam. These tests aimed to report the load–deflection curve, the ultimate flexural capacity, and the interface slip and slip strain values between steel and concrete along the span direction.

The objective of the current paper was to demonstrate a proposed analytical finite element model of continuous double steel-concrete composite beams to estimate the fracture behavior and interface slip values of tested specimens produced by Duan et al.[24], through Ansys 11. The analytical model and the results of system level study can be of interest in assessing progressive collapse resistance of existing structures contain double steel-concrete composite beams and in the design of new structures.

RESEARCH SIGNIFICANCE

The target of this research is to demonstrate a better analytical understanding of double steel-concrete composite beams. Thereby, the focus should be set on the analysis of the maximum increase in strength and deflection capacity due to the existing of double composite action. Therefore, the principal purpose is the nonlinear finite element analysis of continuous steel-concrete composite beams containing double composite action and head studs shear connectors. Within this framework, several aspects should be investigated such as the load–deflection response of the composite beam, and the gradual evolution of slip and slip-strain values at the beam-slap interface up to failure considering double composite action. Based on this investigation, a simplified analytical model through Ansys 11 software is developed in order to enables the prediction of the fracture behavior. Its results are compared with the previously available experimental investigated models introduced by Duan et al. [24]. The results demonstrate a better approximation for the failure criteria in both cases.

METHODOLOGY AND THE ANALYTICAL MODEL

The objective of this section is to describe the finite element model features common to double steel-concrete composite beams being considered. The Ansys 11 finite element package was used to carry out the modeling. The applied load was iterated step by step using the Newton-Raphson method.

Solid65 element was used to model the concrete. This element has eight nodes with three degrees of freedom at each node translations in the nodal x, y, and z directions. The element is capable of plastic deformation, cracking in three orthogonal directions, and crushing. A schematic of the element was shown in Fig. 2a. A Link8 element was used to model steel reinforcement. This element is a 3D spar element and it has two nodes with three degrees of freedom translations in the nodal x, y, and z directions. This element is capable of plastic deformation and element was shown in Fig. 2b. The modeling of the head studs shear connectors was done by the BEAM 188 elements, which allow for the configuration of the cross section, enable consideration of the nonlinearity of the material and include bending stresses. This element was indicated in

Fig. 3a. SOLID185 is used for the modeling of the steel beam. It is defined by eight nodes having three degrees of freedom at each node, translations in the nodal $x, y,$ and z directions. The element has plasticity, hyperelasticity, stress stiffening, creep, large deflection, and large strain capabilities. It also has mixed formulation capability for simulating deformations of nearly incompressible elastoplastic materials, and fully incompressible hyperelastic materials as shown in Fig. 3b. TARGE170 and CONTA173 elements were used to represent the contact slab-steel beam interface. These elements are able to simulate the existence of pressure between them when there is contact, and separation between them when there is not. The two material contacts also take into account friction and cohesion between the parties. The stud shear connector was considered as a clamped metal pin in the steel section, with rotations and translations made compatible. On the slab connector interface, translational referring to the Y and Z axes was also made compatible and, at the Node below the pin head, there was a consideration of coupling in the X direction to represent the mechanical anchoring between the head of the connector and the concrete slab. The geometry of these elements is as shown in Fig. 4a–c. An eight-node solid element, Solid 45, was used to model the steel plates under the load. The element is defined with eight nodes having three degrees of freedom at each node in the nodal $x, y,$ and z directions. The geometry and node locations for this element type are as shown in Fig. 5. Three double steel-concrete composite beam models with the same material properties and cross section shape were analyzed. The only difference between them is that the arrangement of the head studs. Two lines with different number of head studs for the top and the bottom slabs were proposed as reported in Table 1. The geometry of the proposed model components is as shown in Fig. 6a–g. In order to saving Ansys 11 – computational time significantly, a quarter of full composite beams have been modeled as shown in Fig. 7a and b. All the investigated models are constrained at edge ab in the directions y and z, while edge cd is constrained in the directions x and z. In addition, other directions were free of constraints as indicated in Fig. 7a and b. Thus, the research concerns solely symmetrically continuous double steel-concrete composite beams. The cross sections for all the models, namely SCB1, SCB2, and SCB3, are constructed by a top concrete slab along the whole beam length with tension reinforcement 7Φ8/m' in each direction, and by a 1000 mm length bottom concrete slab over interior support, whereas the upper and lower slab thickness was 80 mm.

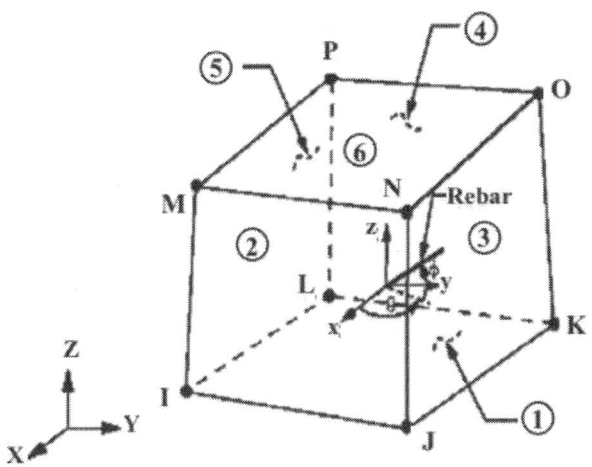

Figure 2a. Solid65 – 3D solids modeling.

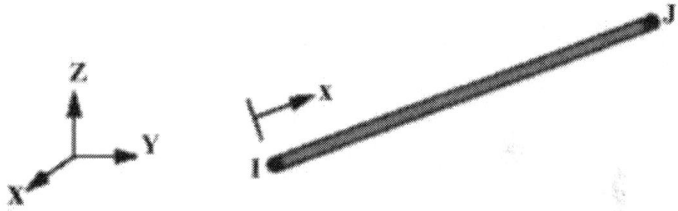

Figure 2b. Link8 –3D spar modeling.

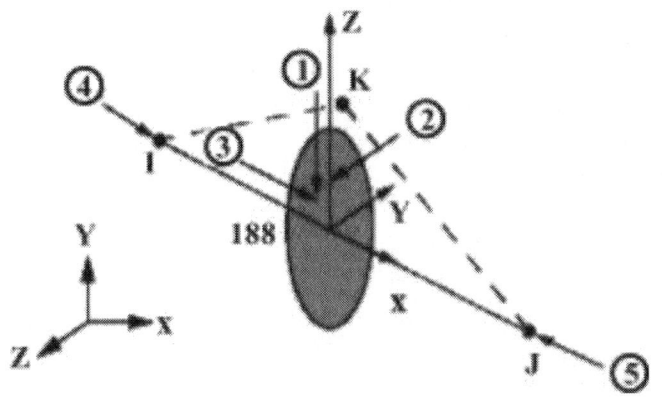

Figure 3a. Beam188 – 3D quadratic beam modeling.

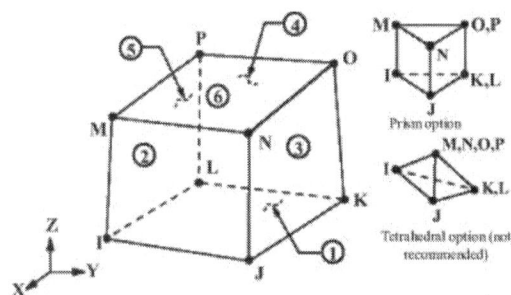

Figure 3b. Solid 185 – 3D solids modeling.

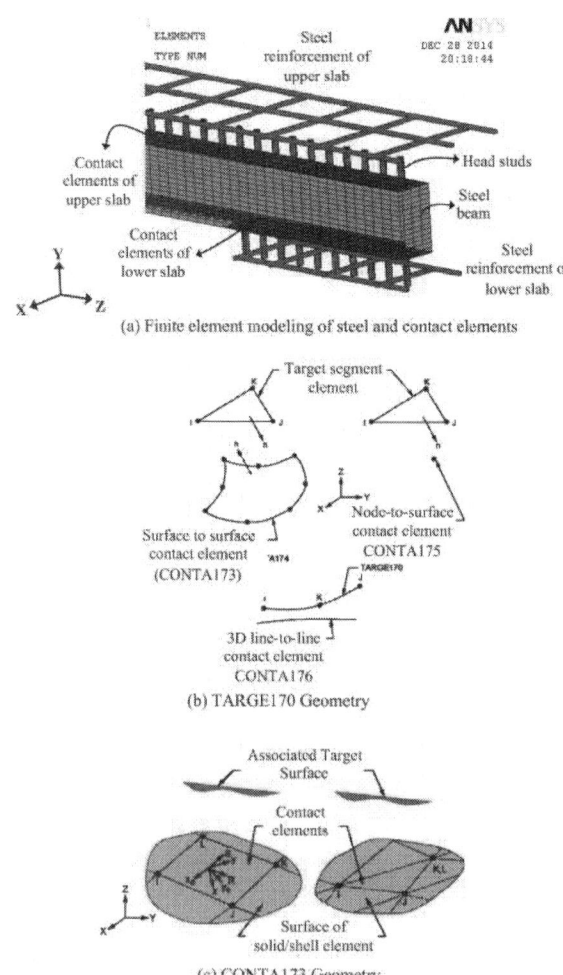

Figure 4. Geometry of TARGE170 and CONTA173 elements.

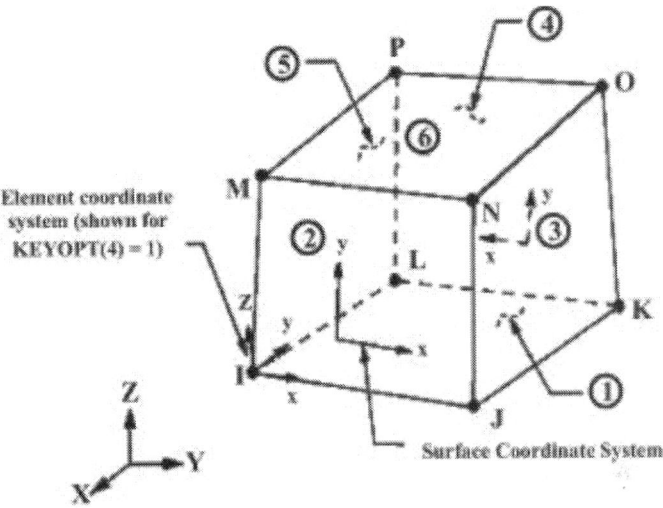

Figure 5. Solid45 – 3D solids modeling.

Table 1. Studs arrangement for the upper and lower slabs.

| Model | Number of studs in each line | |
	Upper slab (N_1)	Lower slab (N_2)
SCB1	94	28
SCB2	82	28
SCB3	82	24

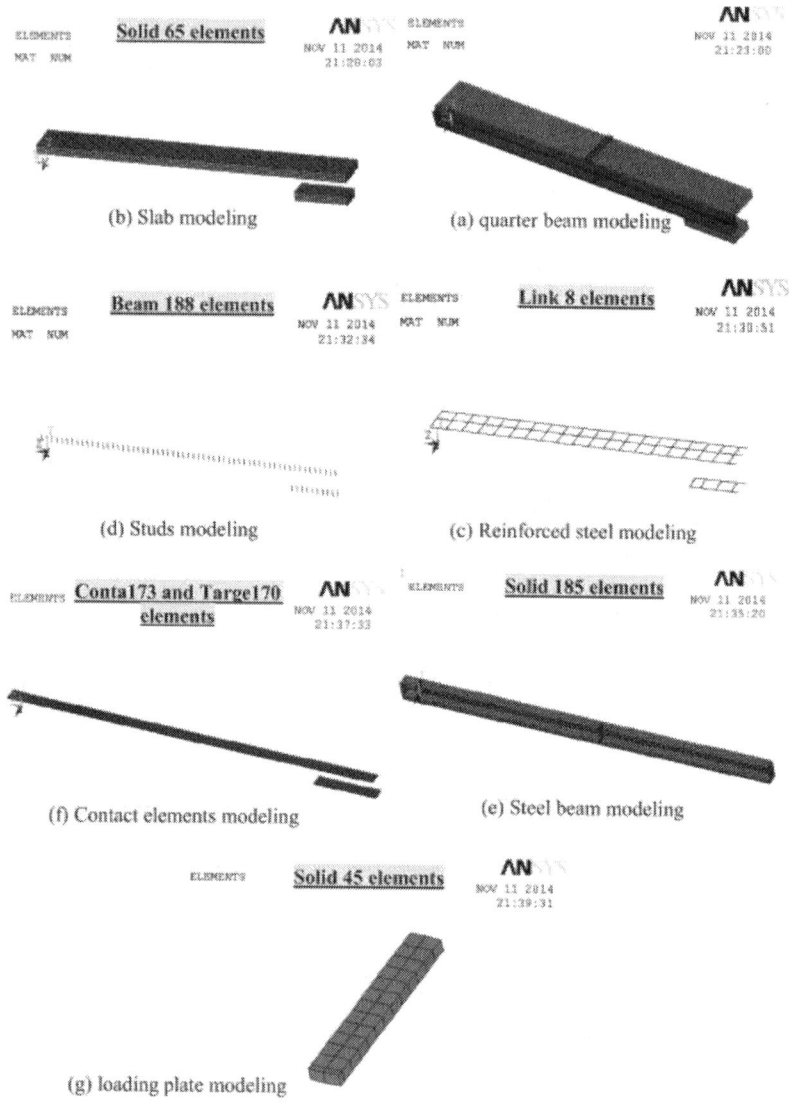

Figure 6. Geometry components of the all beam models.

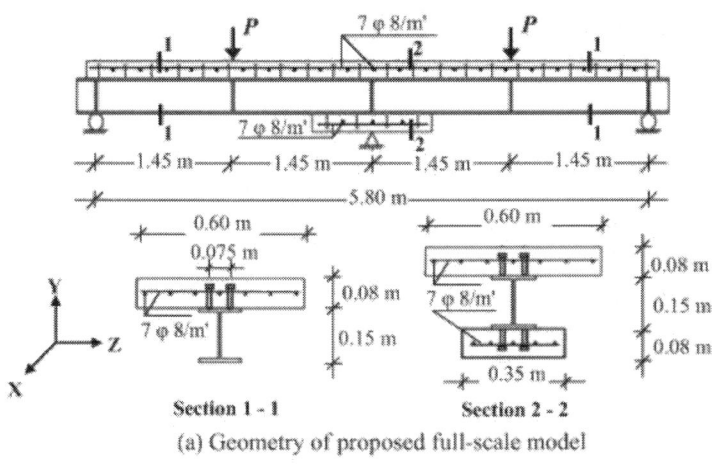

(a) Geometry of proposed full-scale model

(b) Geometry of proposed quarter model

Figure 7. Geometry and cross sections dimensions of all beam models.

MATERIAL PROPERTIES OF THE PROPOSED MODEL

Table 2 summarizes the values of the material properties for all composite beam model components, i.e., reinforced concrete slab, steel beam, and head studs. For the steel beam and head studs, the maximum tensile strength obtained from the experimental test as $f_t = 235$ MPa and the young modulus of elasticity as 2.06×10^5 Mpa. As mentioned above, Solid65 element is used to simulate the concrete. According to Fanning [25], this element requires linear and multilinear isotropic material properties to properly model concrete. For the linear isotropic material, the concrete cube compressive strength obtained from the experimental test as 47 MPa, and the young modulus of elasticity as 4.62×10^4 Mpa. The multilinear isotropic material uses the Von Mises failure criterion along with the William and Warnke [26] model to define the failure of the concrete. A three-dimensional failure surface for concrete is shown in Fig. 8. The most significant non-zero principal stresses are in the x and y directions respectively. Three failure surfaces are shown as the projections on the σ_{xp} - σ_{yp} plane. The mode of failure is the function of the sign of σ_{zp} (principal stress in Z direction). For example, if σ_{xp} and σ_{yp}, both are negative (compressive) and σ_{zp} is slightly positive (tensile), cracking would be predicted in a direction perpendicular to σ_{zp}. However, if σ_{zp} is zero or slightly negative, the material is considered as crushed. Implementation of the William and Warnke [26] material model in Ansys 11 requires different constants that must be defined. Shear behavior of SOLID65 element in Ansys 11 is controlled by two-shear transfer coefficient for open and closed cracks. These coefficients represent conditions at the crack allowing for the possibility of shear sliding across the crack face. A number of preliminary analysis were attempted in this study with various values for the shear transfer coefficients (for open and closed cracks) within the below indicated ranges, but Ansys convergence problems were encountered at the following entering values of the William and Warnke[26] constants:

1. Shear transfer coefficient for open crack was entered as 0.5. Its recommended range is from 0.2 to 0.5 as presented by Razaghi et al. [27].
2. Shear transfer coefficient for closed crack was entered as 1. Its recommended range is from 0.0 (for representing a smooth crack, i.e., complete loss of shear transfer), to 1 (for representing a rough crack, i.e., no loss of shear transfer), as suggested by Razaghi et al. [27].

3. Uniaxial tensile cracking stress which was based upon the modulus of rupture; and was entered as 4.70 Mpa.
4. Uniaxial crushing stress was based on the uniaxial unconfined compressive strength, and was entered as 47.0 Mpa, to turn on the crushing capability of the concrete element as discussed by Kachlakev and Miller [28].
5. Biaxial crushing stress.
6. Ambient hydrostatic stress state for use with constant 7 and 8.
7. Biaxial crushing stress under the ambient hydrostatic stress state (constant 6).
8. Uniaxial crushing stress under the ambient hydrostatic stress state (constant 6).
9. Stiffness multiplier for cracked tensile condition.

Table 2. Material properties of the proposed model.

(1) Concrete	
Concrete strength (f_c)	47 Mpa
Young modulus of elasticity (E_c)	4.62×10^4 Mpa
Poison's ratio (γ)	0.3
(2) Steel	
Maximum tensile strength (f_t)	235 Mpa
Young modulus of elasticity (E_t)	2.06×10^5 Mpa
Poison's ratio (γ)	0.2
(3) Studs	
Maximum tensile strength (f_t)	235 Mpa
Young modulus of elasticity (E_t)	2.06×10^5 Mpa
Diameter (Φ_{stud})	13 mm
Height (h_{stud})	60 mm

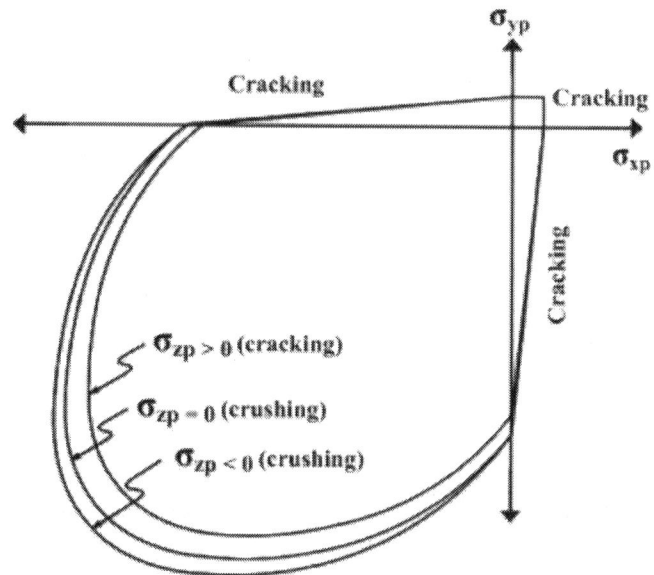

Figure 8. Failure surface for concrete, William and Warnke material model [26].

Coefficients from 5 to 9 were implemented as zero value, as discussed by Wolanski and B. [29], in order to encounter the Ansys convergence problem.

VALIDATION OF THE ANALYTICAL MODEL

The comparison of the results from the analytical model to the experimentally obtained results enables the validation of the performance of the proposed model. The comparison consists of the tests performed by Duan et al. [24] and the results obtained by the proposed finite element model. The proposed model delivered valuable outputs concerning the behavior of the continuous double steel-concrete composite beams such as the strength capacity, the maximum deflection, the interface slip and slip strain of the upper and lower slab of the double composite beam models.

Load–Deflection Relationship

The load–deflection curves analyze the different performance of the double steel-concrete composite model with respect to the strength and deflection capacities. Figure 9, Figure 10 and Figure 11 illustrate the load–deflection curves obtained by both the proposed and experimental approaches for the models SCB1, SCB2, and SCB3 respectively. An increase in the proposed strength capacity values of approximately 32%, 27%, and 29% compared to the experimentally obtained one is observed. Table 3 and Table 4 show the significant comparison of the maximum load capacity and the maximum deflection values for the three proposed models. Good agreement is noticed between the values of the two approaches.

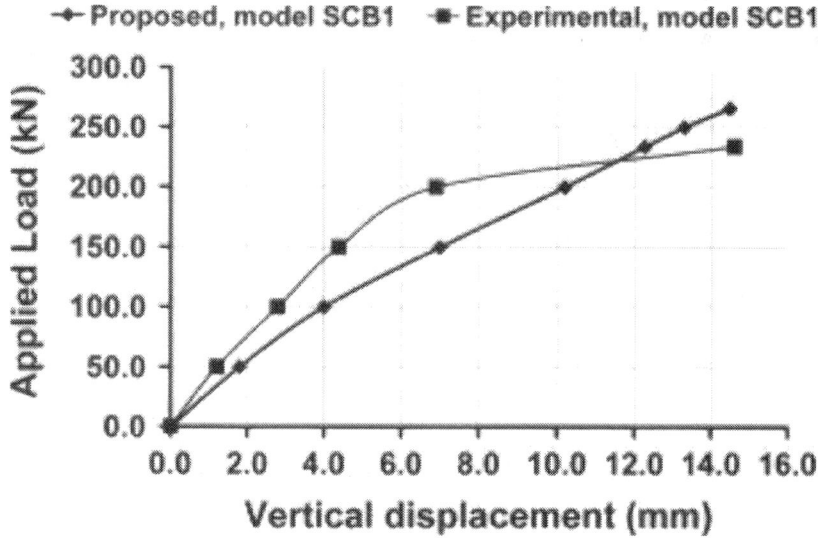

Figure 9. Load verses deflection curve for beam model SCB1.

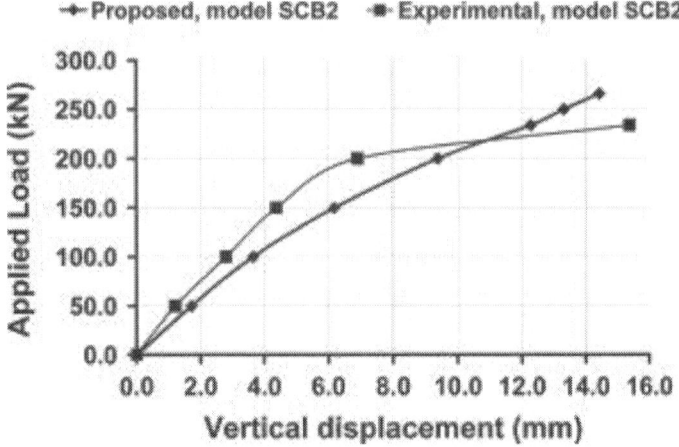

Figure 10. Load verses deflection curve for beam model SCB2.

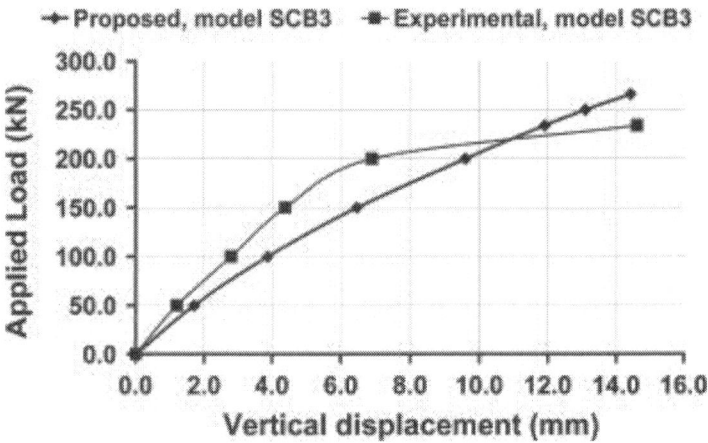

Figure 11. Load verses deflection curve for beam model SCB3.

Table 3. Comparison of the load capacity results at collapse.

| Beam model | Load capacity, $P_{ult.}$ (kN) | | % Difference |
	Proposed	Experimental	
SCB1	266	234	13.67
SCB2	265.5	233	13.73
SCB3	264.3	232	13.79

Table 4. Comparison of the maximum deflection results at collapse.

Beam model	Maximum deflection, $\Delta_{max.}$ (mm)		% Difference
	Proposed	Experimental	
SCB1	14.49	14.61	0.8
SCB2	14.54	15.37	5.4
SCB3	14.43	14.61	1.2

Also, it has to be noticed that the developed models exhibited a softer performance than that of the experimental results. This is due to the following reasons:

1. The William-Warnke failure criteria in Ansys cannot suitably predict the behavior of reinforced concrete structures, as it does not consider the material softening properly due to the varying range of its constants values, such as the shear transfer coefficient for open and closed crack. In addition, for this kind of failure criteria, the crushed elements are removed from the model and that could lead to premature failure, which is not consistent with the real behavior of reinforced concrete structures.

2. Due to the possibility of the inaccuracy in modeling the postyield behavior of steel rebar material, there is somewhat none agreeable between the finite element results and those of experimental results for postyield behavior.

As a result of these two statements, there is disparity between the proposed model results and those of Duan et al. [24] for the pre- and postyield behavior.

Interface Slip Values along the Beam Length

The slip-beam length curves analyze the different performance of the double steel-concrete composite model with respect to the slip values at collapse along the composite beam model length. Figure 12a, Figure 12b, Figure 13a, Figure 13b, Figure 14a and Figure 14b illustrate the slip-beam length curves obtained by both the proposed and experimental approaches for the models SCB1, SCB2, and SCB3 respectively. A reduction in the proposed slip values of approximately 37%, 31%, and 47% compared to the experimentally obtained one is observed for the upper slabs. In contrast, an increase of approximately 21%, 30%, and 28% for the lower slabs is noticed. Good agreement is noticed between the values of the two approaches for the cases of the upper and lower slabs. Fig. 15 shows the steps of the beam-slab interface slip calculation for the upper slab of proposed model SCB1 as an example of the others.

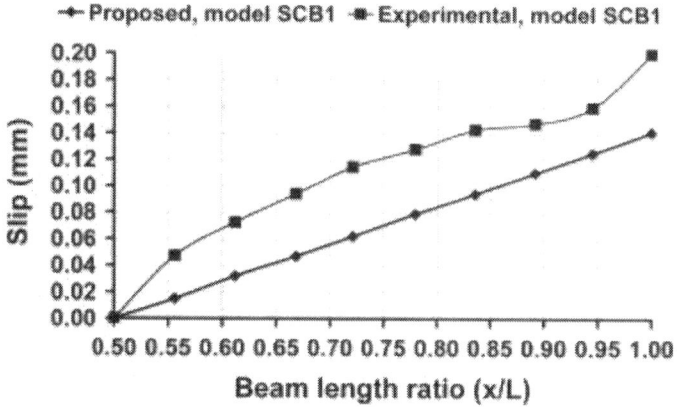

Figure 12a. Interface slip values of the upper slab for beam model SCB1.

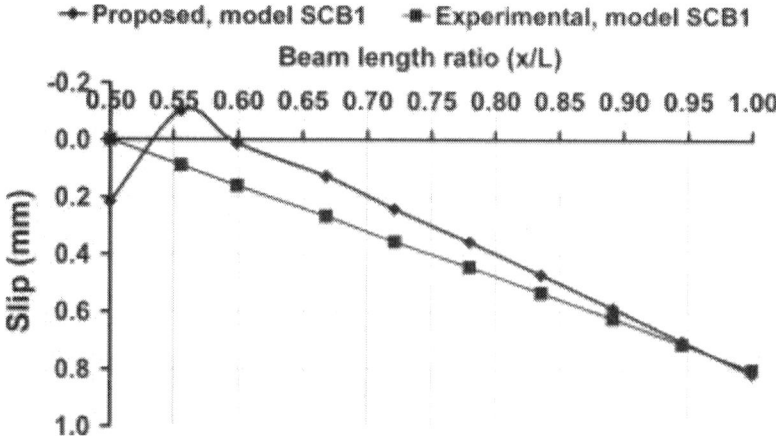

Figure 12b. Interface slip values of the lower slab for beam model SCB1.

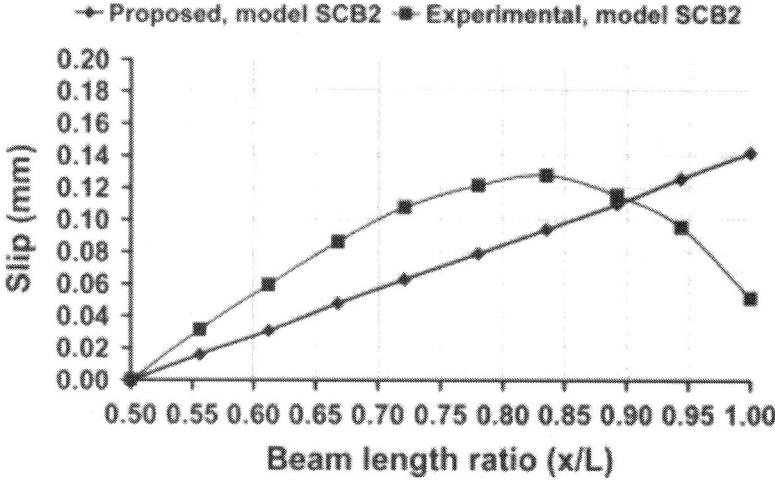

Figure 13a. Interface slip values of the upper slab for beam model SCB2.

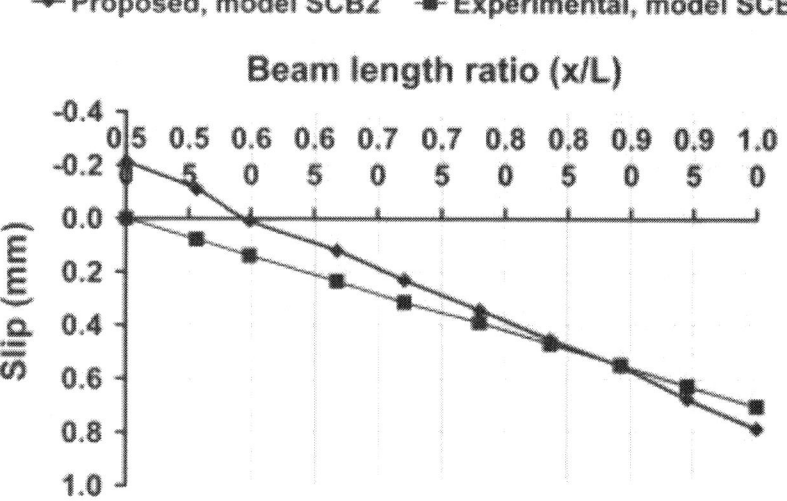

Figure 13b. Interface slip values of the lower slab for beam model SCB2.

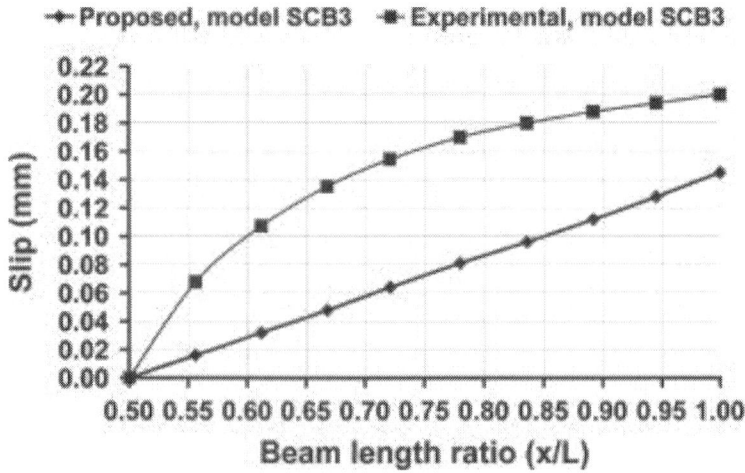

Figure 14a. Interface slip values of the upper slab for beam model SCB3.

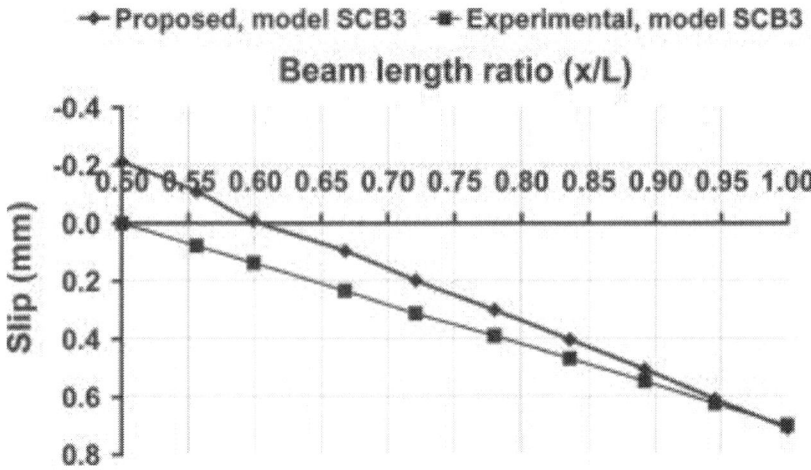

Figure 14b. Interface slip values of the lower slab for beam model SCB3.

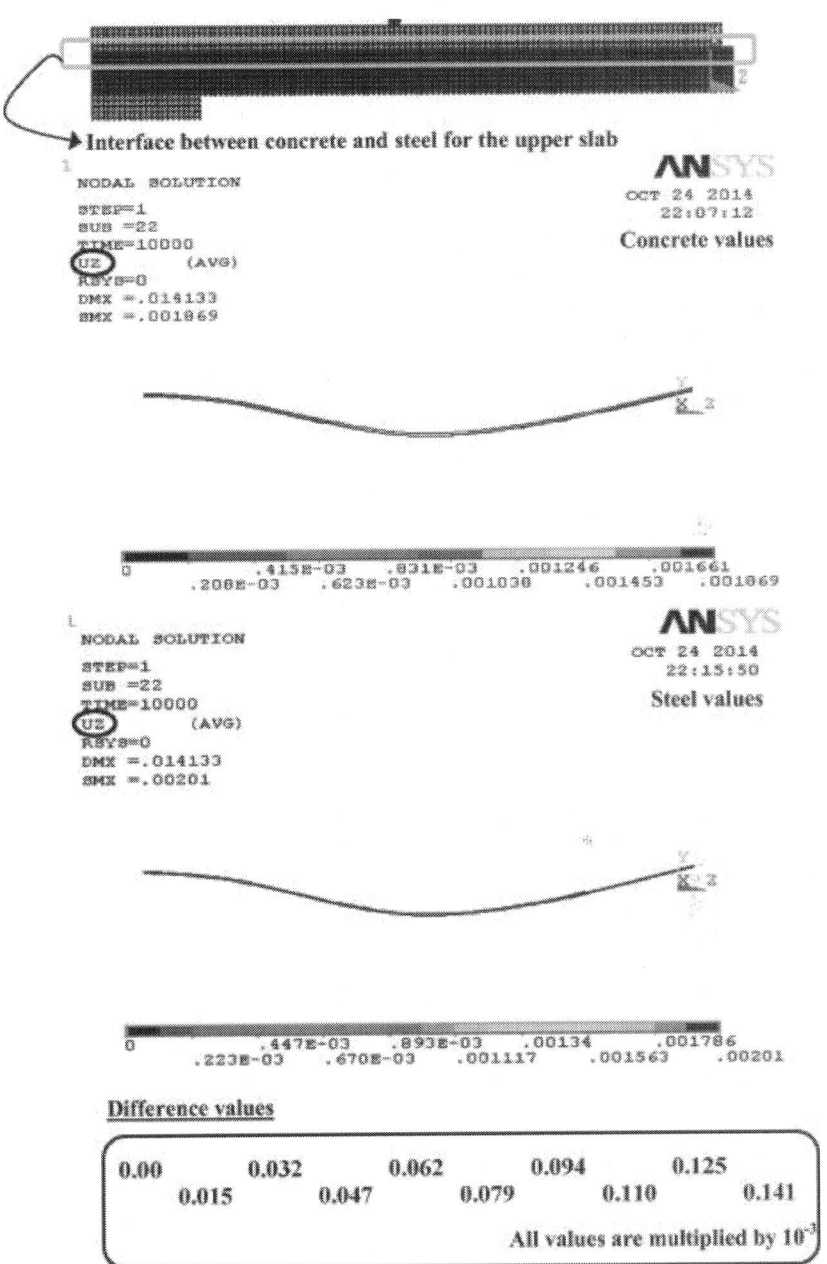

Figure 15. Difference between the interface longitudinal displacements of concrete and steel along the beam length direction for the upper slab of the beam model SCB1 (slip values).

Interface Slip Strain Values along the Beam Length

The slip strain-beam length curves analyze the different performance of the double steel-concrete composite model with respect to the slip strain values at collapse along the composite beam model length. Figure 16a, Figure 16b, Figure 17a, Figure 17b, Figure 18a and Figure 18b illustrate the interface slip strain-beam length curves obtained by both the proposed and experimental approaches for the models SCB1, SCB2, and SCB3 respectively. An increase in the proposed slip strain values of approximately 34%, 52%, and 63% compared to the experimentally obtained one is observed for the upper slabs. In addition, an increase of approximately 35%, 74%, and 62% for the lower slabs is noticed. Somewhat notable non-agreeing values are observed between the values of the two approaches for the cases of the upper and lower slabs. Fig. 19 shows the steps of the interface slip strain calculation for the upper slab of proposed model SCB1 as an example of the others.

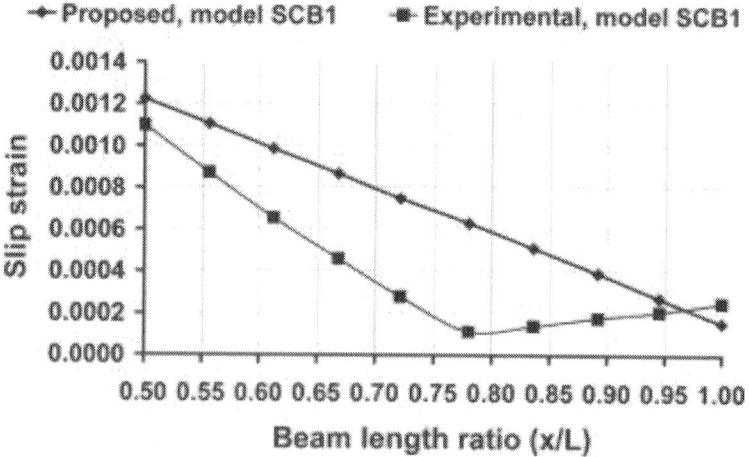

Figure 16a. Interface slip strain values of the upper slab for beam model SCB1.

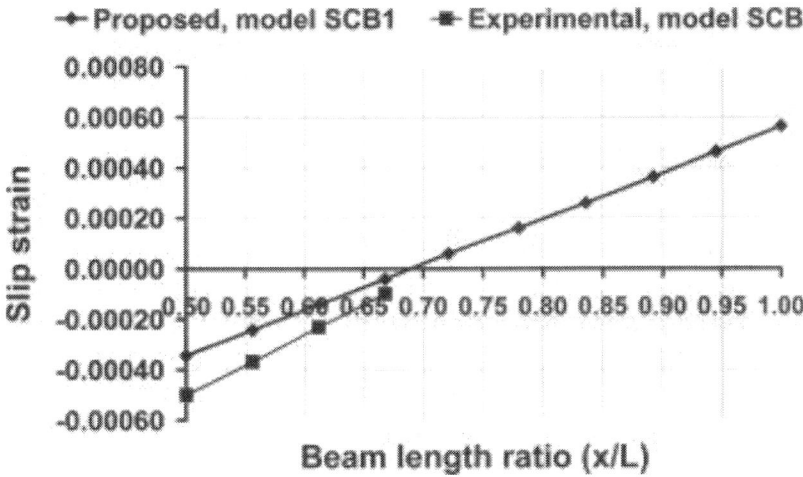

Figure 16b. Interface slip strain values of the lower slab for beam model SCB1.

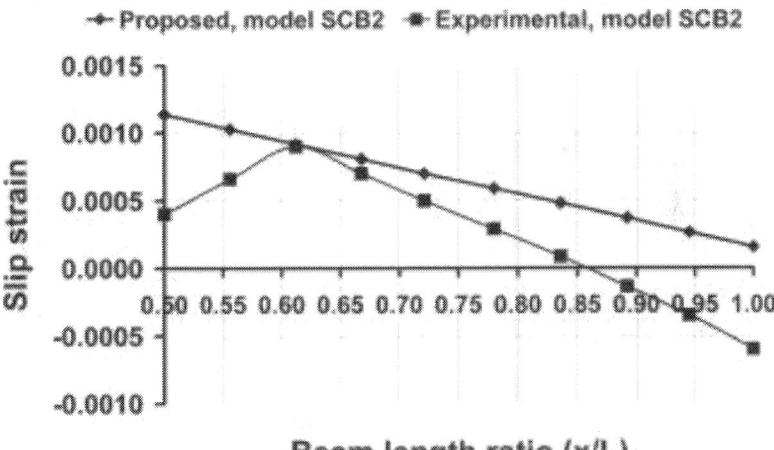

Figure 17a. Interface slip strain values of the upper slab for beam model SCB2.

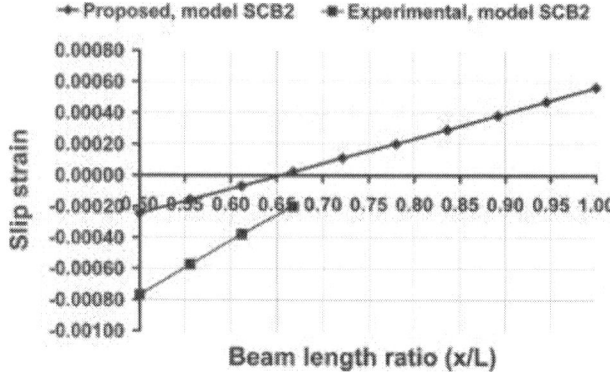

Figure 17b. Interface slip strain values of the lower slab for beam model SCB2.

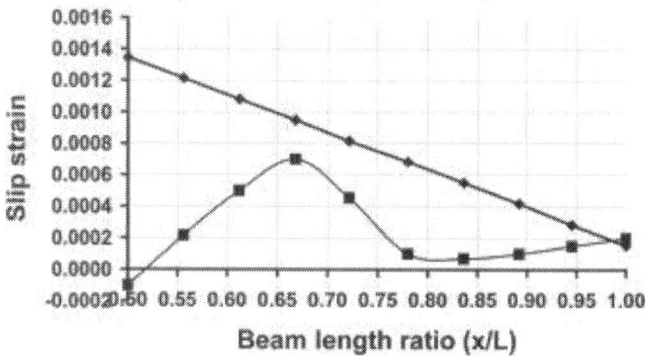

Figure 18a. Interface slip strain values of the upper slab for beam model SCB3.

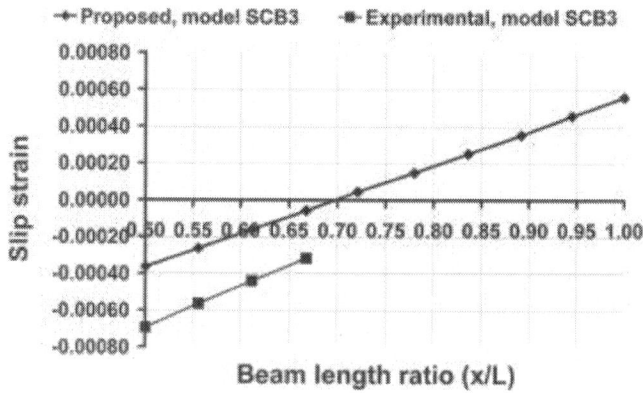

Figure 18b. Interface slip strain values of the lower slab for beam model SCB3.

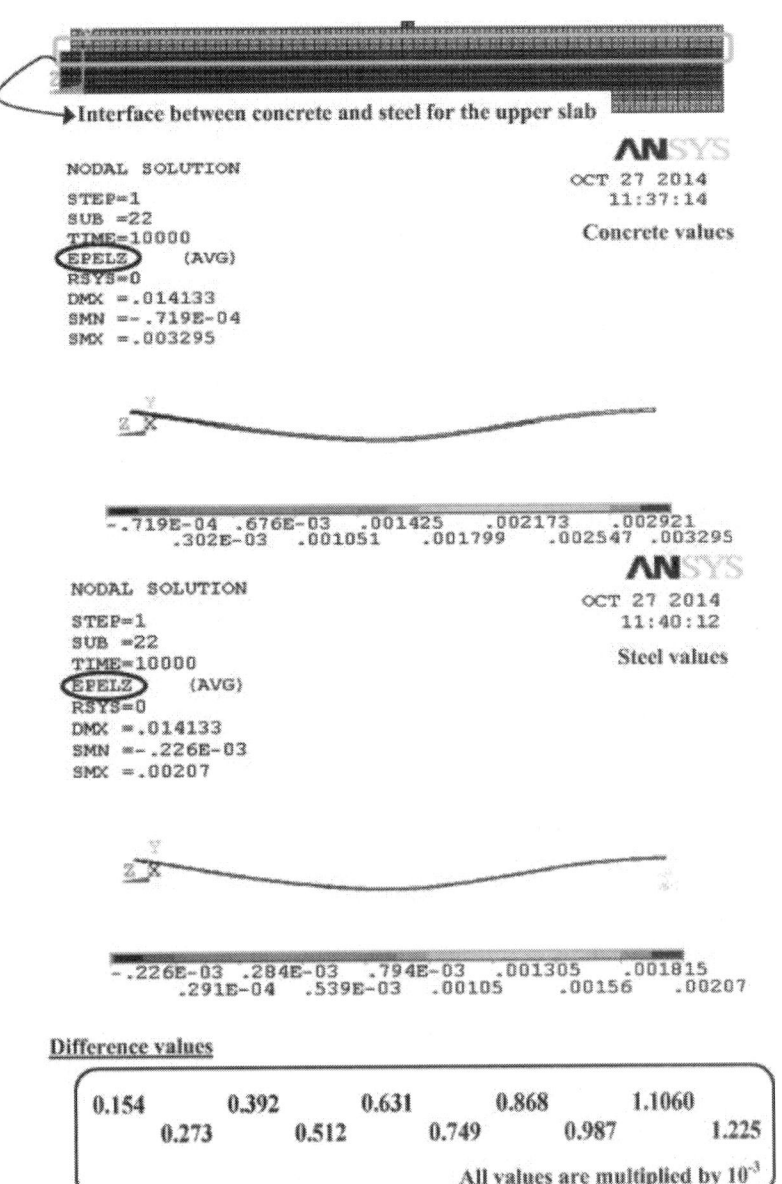

Figure 19. Difference between the interface longitudinal strains of concrete and steel along the beam length direction for the upper slab of the beam model SCB1 (slip strain values).

PARAMETRIC STUDIES

To further improve the understanding of the strength capacity and the fracture behavior of the continuous double steel-concrete composite beams having head studs shear connectors, parametric studies were performed to investigate the impact of the presence or absence of lower slab at the interior support, and the variation of the steel beam height. In addition, the variation of the lower slab length and thickness, and the variation of the studs arrangement and diameter are also studied.

The Influence of Removing the Lower Slab

The case study under consideration involves the influence of removing the lower slab on the mechanical and geometrical characteristics of the beam models at failure, such as the strength and the deflection capacity values. The study was conducted on three proposed models SB1, SCB2, and SCB3 respectively. Fig. 20 illustrates the effect of varying composite action on the fracture characteristics (strength and maximum deflection) of the proposed model.

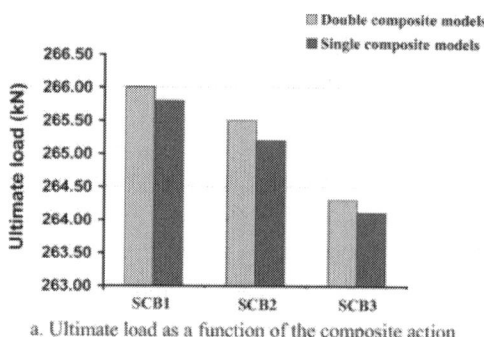

a. Ultimate load as a function of the composite action

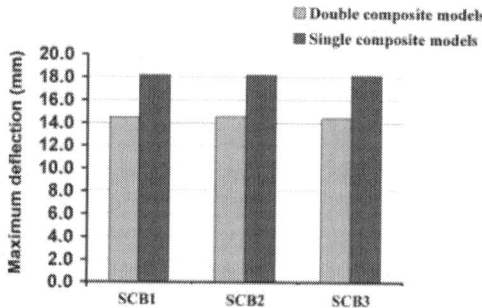

b. Maximum deflection as a function of the composite action

Figure 20. Fracture characteristics as a function varying composite action.

Fig. 20.a compares the results obtained for the ultimate load values, taking into account the presence or absence of the lower slab. It has to be noted that in case of existing the lower slab, the proposed ultimate load values increase by an amount of 0.075% in the case of model SCB1, and by an amount of 0.11% in the case of model SCB2. This increasing value is become 0.068% in the case model SCB3. One can observe that the presence of the lower slab increases the strength capacity by an average amount 0.08% for all experienced composite models. This means that the existence of the lower slab has a minor effect on the strength capacity values.

Fig. 20.b presents a comparison of the results of the maximum deflection values. Again, the above three models were investigated twice in order to experience the effect of removing the lower slab. It has to be noted that in case of existing the lower slab, the proposed values of the maximum deflection decrease by a significant average amount of 20% for all beam models.

The Influence of Varying the Steel Beam Height
In this part, the effect of changing the height of steel beam on the characteristics of the collapse stage for the continuous double steel-concrete composite beam is investigated. Five steel beam heights of values 110 mm, 130 mm, 150 mm, 170 mm, and 190 mm were proposed and applied to the model SCB1, as a case study. Fig. 21 demonstrates the effect of varying steel beam height on the fracture characteristics of the proposed model.

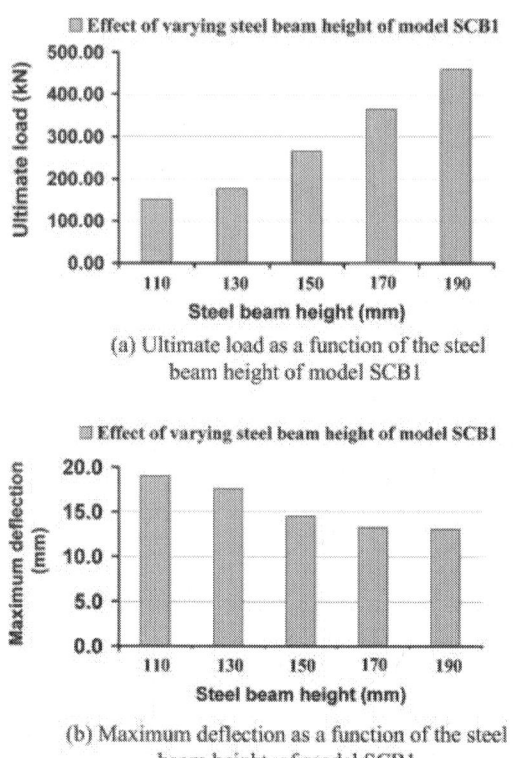

(a) Ultimate load as a function of the steel
beam height of model SCB1

(b) Maximum deflection as a function of the steel
beam height of model SCB1

Figure 21. Fracture characteristics as a function varying steel beam height of model SCB1.

Fig. 21.a presents a comparison of the results of the ultimate load values. This study was applied to the model SCB1 with the same height values as indicated previously. It has to be noted that the case of the beam model with steel beam height of 110 mm had the minimum ultimate load value, whereas the case of the steel beam height of 190 mm had the maximum one. The increase in the ultimate load for two consecutive heights (e.g. 130 mm and 150 mm) reached a significant value of approximately 30% for all models.

Fig. 21.b compares the results obtained for the maximum deflection values, taking into account the same model and the proposed steel beam heights as indicated above. It has to be observed that the case of the 110 mm steel beam height had the maximum value of the maximum deflection, whereas the case of the steel beam with height of 190 mm had the minimum one. The decrease in the maximum deflection values for two consecutive heights (e.g. 130 mm and 150 mm) reached a significant value of approximately 12% for all models.

The Influence of Varying Lower Slab Length

This part contains study of the impact of changing the lower slab length on the collapse stage characteristics for the continuous double steel-concrete composite beam. Four lower slab lengths of values 1000 mm, 1200 mm, 1400 mm, and 1600 mm were proposed and applied to the model SCB1, as a case study. Fig. 22 exhibits the effect of varying the lower slab length on the fracture characteristics of the proposed model.

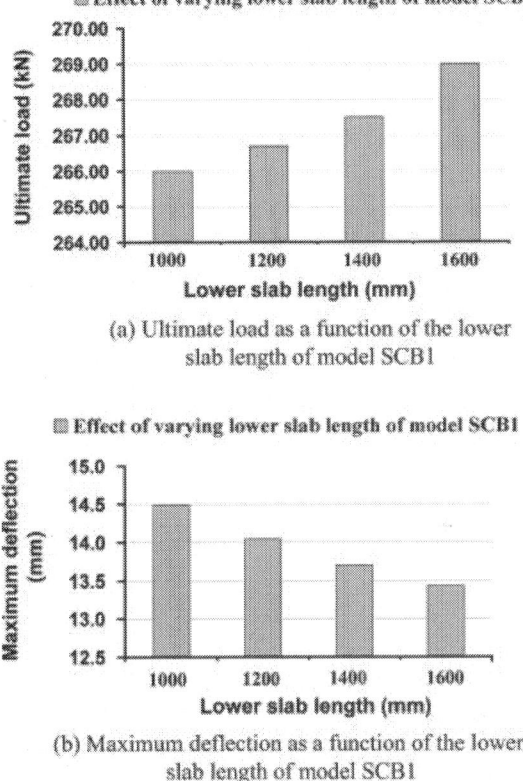

(a) Ultimate load as a function of the lower slab length of model SCB1

(b) Maximum deflection as a function of the lower slab length of model SCB1

Figure 22. Fracture characteristics as a function varying lower slab length of model SCB1.

Fig. 22.a presents a comparison of the results of the ultimate load values. This study was applied to the model SCB1 with the same lower slab length values as mentioned above. It has to be observed that the case of the beam model with lower slab length of 1600 mm had the maximum ultimate load value, whereas the case of the lower slab length of 1000 mm had the minimum one. The increase in the ultimate load for two consecutive slab lengths (e.g. 1200 mm and 1400 mm) reached non-notable value of approximately 0.6% for all models.

Fig. 22.b compares the results obtained for the maximum deflection values, taking into account the same model and the proposed lower slab lengths as indicated above. It has to be noted that the case of the beam model involving lower slab length of 1600 mm had the minimum value of the maximum deflection, whereas the case of the lower slab length of 1000 mm had the maximum one. The decrease in the maximum deflection values for two consecutive slab lengths (e.g. 1200 mm and 1400 mm) reached a remarkable value of approximately 5% for all models.

The Influence of Varying Lower Slab Thickness

The influence of changing the lower slab thickness on the characteristics of the collapse stage for the continuous double steel-concrete composite beam is studied herein. Four lower slab thicknesses of values 80 mm, 100 mm, 120 mm, and 140 mm were proposed and executed to the model SCB1, as a case study. Fig. 23 explicates the effect of varying the lower slab thickness on the fracture characteristics of the proposed model.

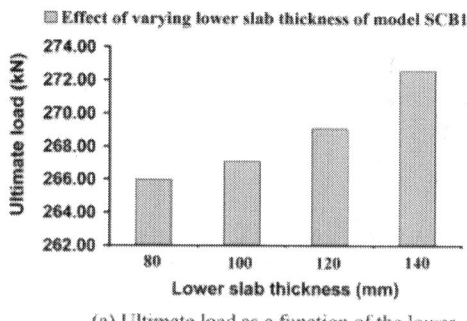

(a) Ultimate load as a function of the lower slab thickness of model SCB1

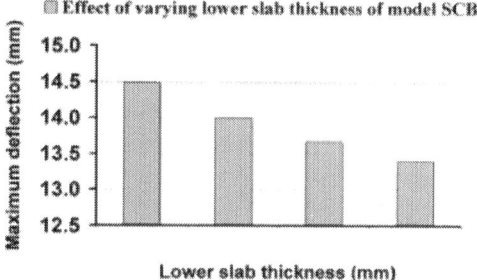

(b) Maximum deflections as a function of the lower slab thickness of model SCB1

Figure 23. Fracture characteristics as a function varying lower slab thickness of model SCB1.

Fig. 23.a presents a comparison of the results of the ultimate load values. This study was applied to the model SCB1 with the same lower slab thickness values as mentioned above. It has to be noted that the case of the beam model with lower slab thickness of 80 mm had the minimum ultimate load value, whereas the case of the lower slab thickness of 140 mm had the maximum one. The increase in the ultimate load for two consecutive slab thicknesses (e.g. 100 mm and 120 mm) reached non-remarkable value of approximately 0.85% for all models.

Fig. 23.b compares the results obtained for the maximum deflection values, taking into account the same model and the proposed lower slab thicknesses as mentioned above. It has to be observed that the case of the beam model including lower slab thickness of 80 mm had the maximum value of the maximum deflection, whereas the case of the lower slab thickness of 140 mm had the minimum one. The decrease in the maximum deflection values for two consecutive slab thicknesses (e.g. 100 mm and 120 mm) reached a slightly remarkable average value of approximately 3.5% for all models.

The Influence of Varying the Head Studs Arrangement

In order to complete the parametric study, the effect of changing the arrangement of the head studs on the characteristics of the collapse stage for the continuous double steel-concrete composite beam is discussed. Three cases of head studs arrangement were proposed. The first case is when the studs were fully arranged along the whole length of the upper and the lower interface slab-steel beam surfaces. The second case is when the studs were arranged with staggered shape, while the third case is when the studs were completely removed. This study was applied to the model SCB1, as a case study. Fig. 24indicates the effect of varying the studs arrangement on the fracture characteristics of the proposed model.

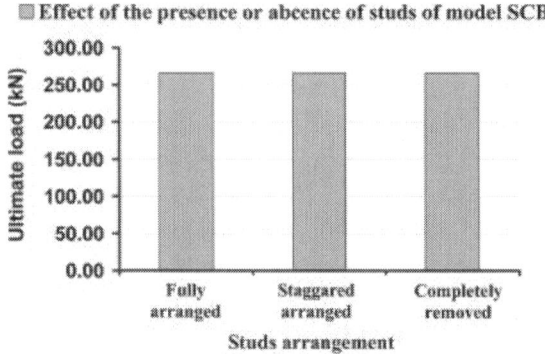

(a) Ultimate load as a function of the studs
arrangement of model SCB1

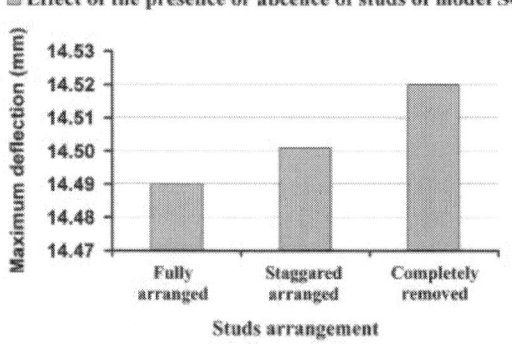

(b) Maximum deflection as a function of the studs
arrangement of model SCB1

Figure 24. Fracture characteristics as a function varying studs arrangement of model SCB1.

Fig. 24.a presents a comparison of the results of the ultimate load values. This study was applied to the model SCB1 with the same cases of the studs arrangement as mentioned above. It has to be noted that the change of the shape of the studs arrangement has no influence on values of the ultimate load.

Fig. 24.b compares the results obtained for the maximum deflection values. It has to be observed that the beam model including fully studs arrangement had the minimum value of the maximum deflection, whereas the case of the completely removed head studs had the maximum one. The increase in the maximum deflection values for two consecutive studs arrangement (e.g. fully and staggered arrangement) reached a very slightly remarkable value of approximately 0.08% for all models.

The Influence of Varying the Head Studs Diameter

The effect of changing the value of the diameter of the head studs on the characteristics of the collapse stage for the continuous double steel-concrete composite beam is examined as a part of this study. Four cases of head studs diameter of values 13 mm, 16 mm, 19 mm, and 22 mm were suggested and implemented to the model SCB1, as a case study. Fig. 25 clarifies the effect of varying the studs diameter on the fracture characteristics of the proposed model.

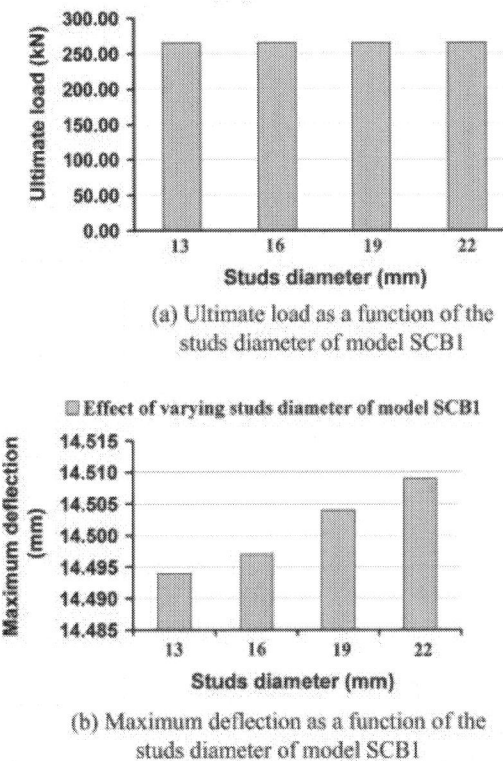

(a) Ultimate load as a function of the studs diameter of model SCB1

(b) Maximum deflection as a function of the studs diameter of model SCB1

Figure 25. Fracture characteristics as a function varying studs diameter of model SCB1.

Fig. 25.a presents a comparison of the results of the ultimate load values. This study was applied to the model SCB1 with the same values of the head studs diameters as stated above. It has to be noted that the change of the head studs diameter has no influence on values of the ultimate load.

Fig. 25.b compares the results obtained for the maximum deflection values. It has to be concluded that the beam model containing 13 mm head studs diameter had the minimum value of the maximum deflection, whereas the case of the studs diameter of 22 mm had the maximum one. The increase in the maximum deflection values for two consecutive studs diameter (e.g. 16 mm and 19 mm) attained a very slightly notable average value of approximately 0.05% for all models.

CONCLUSIONS

This paper investigates the behavior of the continuous steel-concrete composite beam taking into account the existence of the double composite action and the head stud shear connectors.

Based on the finite element numerical study and the experimentally available results, the following main conclusions can be extrapolated:

1. A numerical proposed model based on the finite element theory can be used to examine the geometrical and mechanical characteristics in steel-concrete composite beam with double composite action, resulting in a good agreement when comparing to available full-scale test data.
2. The comparison of the strength capacity values obtained by the proposed and experimental models leads to a good agreeable between them. An average increase in the proposed strength capacity values of approximately 29% compared to the experimentally available data was concluded for all proposed models. However, a softer performance of the validation figures (load – deflection curves) is observed for the developed models than that of the experimental results. This is mainly due to the varying range of the William-Warnke constants values, which must be chosen carefully by a sensitivity analysis in order to encounter the Ansys convergence problems as mentioned above.
3. An increase in the proposed interface steel-concrete slip values of approximately 38% compared to the experimentally available data was observed, leading to slightly non-agreeable results.
4. An increase in the proposed interface steel-concrete slip strain values of approximately 49% and 55% compared to the experimentally available data was observed for both the upper and the lower slabs respectively, leading to somewhat non-agreeing values between them. This is due to the difference in values of friction (shear slip) at the slab-steel beam interface between the analytical and experimental approaches, because of the presence of the contact elements for

simulating this friction. This means that the shear slip has a significant contribution to composite beam deformation, which cannot be negligible.

5. Parametric studies were carried out to look at the impact of removing the lower slab, the effect of varying steel beam height and the lower slab length and thickness, and the effect of changing the head studs arrangement and diameter. These studies were performed to investigate the effect of these parameters on the strength and the deflection capacity of the steel-concrete composite beams having double composite action.

6. The presence of the lower slab increases the proposed strength capacity values by an average amount 0.08% for all experienced composite models, leading to a minor effect on the strength capacity. Moreover, the proposed values of the maximum deflection decrease by a significant average amount of 20% for all beam models when removing the lower slab.

7. In comparison with the five suggested cases of steel beam height involved in the parametric study, it can be observed that the more increase the steel beam height is the bigger the ultimate load values are.

8. Moreover, this study showed that the smaller the lower slab length or thickness is the smaller the ultimate load values and the bigger the maximum deflection values are.

9. It can be noted that the change of the shape of the studs arrangement has no influence on the values of the ultimate load. In addition, the beam model including fully studs arrangement had a minimum value of the maximum deflection, whereas the case of the completely removed head studs had the maximum one.

10. In comparison with the five head stud diameters suggested in this study, one can concluded that the change of this parameter has no effect on the values of the ultimate load. In addition, it has to be noted that the smaller the head studs diameter is the smaller the maximum deflection values are.

REFERENCES

1. Rozsas A. Plastic design of steel–concrete composite girder bridges. M.sc thesis, department of structural engineering, faculty of civil engineering: Budapest (Hungary); 2011.

2. Xu C, Su Q, Wu C. Experimental study on double composite action in the negative flexural region of two-span continuous composite box girder. J Constr Steel Res 2011;67(10):1636–48.

3. Tan El, Uy B, Hummam G. Behavior of multi-span composite steel–concrete beams subjected to combined flexure and torsion. Research and applications in structural engineering, mechanics and computation, London, UK; 2013. p. 1397–402.

4. Lin W, Yoda T. Numerical study on horizontally curved steel– concrete composite beams subjected to hogging moment. Int J Steel Struct, USA 2014;14(3):557–69.

5. Henriques D, Goncalves R, Gamotim D. Nonlinear analysis of steel–concrete beams using generalized beam theory. In: 11th World Congress on Computational Mechanic, Barcelona, Spain; 20–25 July 2014.

6. Liang Q, Uy B, Bradford M, Ronapf H. Ultimate strength of continuous composite beams in combined bending and shear. J Constr Steel Res 2004;60(8):1109–28.

7. Liang Q, Uy B, Bradford M, Ronapf H. Strength analysis of steel–concrete composite beams in combined bending and shear. J Struct Eng, ASCE, USA 2005;131(10):1593–600.

8. Sebastian W, McConnel RE. Nonlinear finite element analysis of steel–concrete composite structures. J Struct Eng, ASCE, USA 2000;126(6):662–74.

9. Hirst MJS, Yeo MF. The analysis of composite beams using standard finite element programs. Comput Struct 1980;11(3):233–7.

10. Al-Amery RIM, Roberts TM. Nonlinear finite difference analysis of composite beams with partial interaction. Comput Struct 1990;35(1):81–7.

11. Salari MR, Spacone E, Shing B, Frangopol DM. Nonlinear analysis of composite beams with deformable shear connectors. J Struct Eng, USA 1998;124(10):1148–58.

12. Thevendran V, Chen S, Shanmungam NE, Liew JWR. Nonlinear analysis of steel–concrete composite beams curved in plan. Finite Elem Anal Des 1999;32(3):125–39.

13. Reiner S. Bridges with double composite action. Struct Eng Int, UK 1999;1:32–6.

14. Stroh SL, Sen R.** Steel bridge with double composite action: innovative design. In: 5th International bridge engineering conference, tampa, FL (US). Transportation Research Record, April 3–5, 2000. vol. 1, 1696. p. 299–309.

15. Newmark NM, Siess CP, Viest IM. Test and analysis of composite beams with incomplete interaction. Proc, Soc Exp Stress Anal 1951;9:75–92.

16. Duan S, Niu R, Xu J, Zheng H. A finite element model for double composite beam. Challenges, opportunities and solutions in structural engineering and construction, London; 2010. p. 197– 202.

17. Duan SJ, Huo JH, Zhou QD. The research on calculation method of the ultimate bearing capacity of double steel– concrete composite beam. J Shijiazhuang Railway Inst 2007;20(4):1–4.

18. Duan SJ, Zhou QD, et al. Experimental study on bearing capacity of double steel and concrete composite continuous beams. J Railway Sci Eng 2008;5(5):12–7.

19. Nagai M, Inaba N, et al. Experimental study on ultimate strength of composite and double composite girders. In: Proceedings of 8th Pacific structural steel conference steel structures in natural hazards, 2007. p. 329– 34.

20. Duan SJ, Duan YJ, Zhang ZG. The interface slip expression of double steel– concrete composite beam under concentrated load. J Shijiazhuang Railway Inst 2007;20(2):1–4.

21. Yang XW, Duan SJ. The effective width of reinforcement bars for double steel–concrete composite beam. Eng Mech 2008;25(A1): 184–8.

22. Wang G, Wang FJ, et al. Theoretical analysis of double composite beam deformation in elastic state by Goodman elastic sandwich method. Chin Railway Sci 2006;27(5):66–70.

23. Yen BT, Huang T, et al. Steel box girders with composite bottom flanges. In: Official proceedings, 3rd annual international bridge conference. Pittsburgh, PA (US); 1986. p. 79–86.

24. Duan SJ, Wang JW, Zhou QD, Wang HL. An experimental study on double steel–concrete composite beam specimens. Challenges, opportunities and solutions in structural engineering and construction, London; 2010. p. 209– 14.

25. Fanning P. Nonlinear models of reinforced and post tensioned concrete beams. Lecture, Department of Civil Engineering University College Dublin Earls fort Terrace. Dublin, Ireland; 2001.

26. William KJ, Warnke ED. Constitutive model for the triaxial behavior of concrete. Proc of the Int Assoc Bridge Structural Engineering, ISMES, Bergamo 1975;19:174.

27. Razaghi J, Hosseini A, Hatami F. Finite element method application in nonlinear analysis of reinforced concrete structures. Second Nat Congr Civil Eng 2005.

28. Kachlakev D, Miller T. Finite element modeling of reinforced concrete structures strengthened with FRP laminates. Oregon state University; May 2001.

29. Wolanski J, B.S. Flexural behavior of reinforced and prestressed concrete beams using FRP element Analysis, A thesis submitted to the faculty of the graduated school, Marquetee university, in partial fulfillment of the requirement for the degree of master of science; May 2004.

CITATION

Ashraf Mohamed Mahmoud, Finite element modeling of steel concrete beam considering double composite action, Ain Shams Engineering Journal, Available online 4 May 2015, ISSN 2090-4479, http://dx.doi.org/10. 1016/j. asej.2015. 03.012.

CHAPTER 8

Effects of Sand Quality on Compressive Strength of Concrete: A Case of Nairobi County and Its Environs, Kenya

Hannah Nyambara Ngugi[1], Raphael Ndisya Mutuku[2], Zachary Abiero Gariy[2]

[1]Pan African University, Institute for Basic Sciences, Technology and Innovations (PAUISTI) hosted at Jomo Kenyatta University of Agriculture and Technology (JKUAT), Juja, Kenya
[2]Department of Civil, Construction and Environmental Engineering, Jomo Kenyatta University of Agriculture and Technology (JKUAT), Juja, Kenya

ABSTRACT

Failure of concrete structures leading to collapse of buildings has initiated various researches on the quality of construction materials. Collapse of buildings resulting to injuries, loss of lives and investments has been largely attributed to use of poor quality concrete ingredients. Information on the effect of silt and clay content and organic impurities present in building sand being supplied in Nairobi County and its environs as well as their effect to the compressive strength of concrete was lacking. The objective of this research was to establish level of silt, clay and organic impurities present in building sand and its effect on compressive strength of concrete. This paper presents the findings on the quality of building sand as sourced from eight supply points in Nairobi County and its environs and the effects of these sand impurities to the compressive strength of concrete. 27 sand samples were tested for silt and clay contents and organic impurities in accordance with BS 882 and ASTM C40 respectively after which 13 sand samples with varying level of impurities were selected for casting of concrete cubes. 150 mm × 150 mm × 150 mm concrete cubes were cast using concrete mix of 1:1.5:3:0.57 (cement: sand: coarse aggregates: water) and were tested for compressive strength at the age of 7, 14 and 28 days. The investigation used cement, coarse aggregates (crushed stones) and water of similar characteristics while sand used had varying levels of impurities and particle shapes and texture. The results of the investigations showed that 86.2% of the sand samples tested exceeded the allowable limit of silt and clay content of while 77% exceeded the organic content limit. The level of silt and clay content ranged from 42% to 3.3% for while organic impurities ranged from 0.029 to 0.738 photometric ohms for the unwashed sand samples. With regard to compressive strength, 38% of the concrete cubes made from sand with varying sand impurities failed to meet the design strength of 25 Mpa at the age of 28 days. A combined regression equation of $Fcu28 = -23.20SCI - 2.416ORG + 25.57$ 2 8 /> with R2 = 0.444 was generated predicting compressive strength varying levels of silt and clay impurities (SCI), and organic impurities (ORG) in sand. This implies that 44% of concrete's compressive strength is contributed by combination of silt and clay content and organic impurities in sand. Other factors such as particle shapes, texture, workability and mode of sand formation also play a key role in determination of concrete strength. It is concluded that sand found in Nairobi County and its environs contain silt and clay content and organic impurities that exceed the allowable limits and these impurities result in significant reduction in concrete's compressive strength. It is recommended that the concrete design mix should always consider the strength reduction due to presence of these impurities to ensure that target strength of the resultant concrete is achieved. Formulation of policies governing monitoring of quality of building sand in Kenya and other developed countries is recommended.

INTRODUCTION

Quality of constituent materials used in the preparation of concrete plays a paramount role in the development of both physical and strength properties of the resultant concrete. Water, cement, fine aggregates, coarse aggregates and any admixtures used should be free from harmful impurities that negatively impact on the properties of hardened concrete. Sand is one of the normal natural fine aggregates used in concrete production [1]. Past researches identify the major causes of buildings failure as dependent on the quality of building materials used (sand, coarse aggregates, steel reinforcement, water), workmanship employed in the concrete mix proportioning and construction methodology, defective designs and non-compliance with specifications or standards[2] -[7] . This investigation focuses on the quality of building sand in terms of having the silt and clay content and organic impurities within the allowable limit as set out in British Standard (BS) 882.

Quality assurance of building materials is very essential in order to build strong, durable and cost effective structures [8] . When construction is planned, building materials should be selected to fulfill the functions expected from them. In Kenya over 14 buildings have been reported to collapse in the last 10 years leading to deaths and injuries (see Appendix 1) and various cause of building failure have been suggested. Use of poor quality construction materials (such as quality of sand, aggregates or water) result in poor quality structures and may cause structures to fail leading to injuries, deaths and loss of investment for developers. Impurities in building sands contribute to reduced compressive strength. Olanitori [9] asserts that the higher the percentage of clay and silt content in sand used in concrete production, the lower the compressive strength of the hardened concrete. Although many studies mentioned above have shown that use of poor quality materials is one of the major contributing factors to collapse of buildings, testing these materials has not been carried out to examine the impact of impurities in building sands to the overall performance of concrete. In addition, where tests have been carried out [10] , testing of both clayey, silts and organic impurities has not been carried out to determine their combined effect on the concrete strength. To prevent buildings failure, careful selection of construction materials including building sands is paramount to ensure they meet the set construction standards. Impurities in sand impact negatively on compressive strength as well as bond strength between steel reinforcement and concrete and may cause buildings failure. BS 882 [11] specifies the tests for suitable aggregates.

The Nigerian Standard Organization specifies the maximum quantity of silt in sand as 8% beyond which sand is regarded as unsuitable for construction work [10] . American Society for Testing and Materials (ASTM) C 117 [12] and Hong Kong [13] construction standards give an allowable limit of 10% for silt and clay content in sand. On the other hand BS 882 states that the percentage of clay and fine silts must not exceed 4% by weight for sand for use in concrete production [14] . Fine aggregates containing more than the allowable percentages of silt are required to be washed so as to bring the silt content within allowable limits. As a thumb rule, the total amount of deleterious materials in a given aggregate should not exceed 5% [15] . The methods of determining the content of these deleterious materials are prescribed by IS 383 [16] , BS 882 [11] , ASTM C 117[12] and [17] . These include determination of contents organic impurities, clay, or any deleterious material or excessive fillers of sizes smaller than No. 100 sieve. This research also seeks to determine the level of silt and clay content and organic impurities present in building sand being supplied in Nairobi County and its environs and also the effect of these impurities to the compressive strength of concrete. It further seeks to establish the minimum allowable limits of silt and clay and organic impurities for concrete production based on the tested samples.

MATERIALS AND METHODS

Materials

The research employed laboratory experimental methods. Sand samples were collected from eight main sand supply points in Nairobi City County and its environs namely Njiru, Mlolongo, Kitengela, Kawangare, Dagoretti Corner, Kariobangi, Kiambu and Thika in Kenya as shown in Figure 1.

Sand samples were labeled based on their point collection where NR , ML, KT, KW, DC, KB, KBU and TK was used to represent samples sourced from Njiru, Mlolongo, Kitengela, Kawangare, Dagoretti Corner, Kariobangi, Kiambu and Thika respectively. A digit number was further given to represent the sample number as collected from each supply point e.g. NR1, NR1, NR3 was used to label Njiru sample 1, Njiru sample 2 and Njiru sample 3 respectively for sand samples sourced from Njiru area. Two sand samples CL1 and CL2 were washed and used as control samples.

From each supply point, 50 kg the selected sand samples were procured for grading and testing. Course aggregates from crushed stones, ordinary Portland cement grade 32.5 and 12 mm diameter twisted steel reinforcement bars were sourced from local manufacturers in Kenya. Clean portable water from the University (JKUAT, Kenya) was used.

Methods

Sieve analysis was carried out on the sand samples to determine their degrees of fineness (seeFigure 2). Percentages of sand passing and retained was analyzed and grading curved plotted for comparison. Control sand sample was prepared by thoroughly washing river sand with clean water to remove silt and clay and organic impurities present and dried. Physical examination of sand particle shapes and sizes was done as well as determination of specific gravity of sand using pycnometer glass vessel as detailed in the IS standard [16] equivalent to ASTM D854 [18] . Sand samples were further tested using Laser Diffraction and Particle Size Analysis (LDPSA) and Total X-ray fluorescence (TXRF) methods to determine the constituent chemical elements and results are shown in Appendix 2. From the preliminary test results on sand impurities found in 27 sand samples, thirteen sand samples were carefully selected for preparation of concrete cube for compressive strength testing in a bid ensure fair distribution and representativeness of all sand sample categories. Figure 2 and Figure 3 below show part of the organic impurities, and silt and clay content testing processes.

Table 1 Summary of sieve analysis and geological grading zones classification.

Grading zones	Frequency (no)	Dominance (%)	Description
Zone I	2	7%	Course sand
Zone II	18	67%	Normal sand
Zone III	5	19%	Fine sand
Zone III	2	7%	Very fine sand

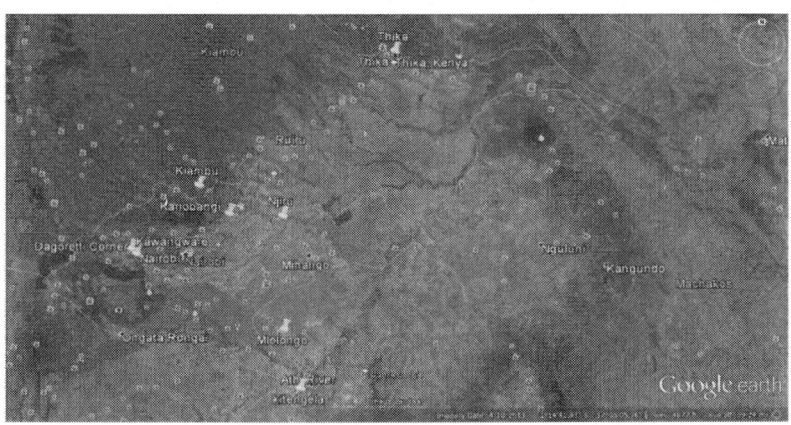

Figure 1. Main sand supply points in Nairobi City County and its environs (source: google earth, January 2014).

Figure 2. Organic impurities testing.

Figure 3. Silt and clay content testing.

Concrete mix ratio of 1:1.5:3:0.57 (cement: sand: coarse aggregates: water) as it is used for most low rise structural buildings was designed for an expected compressive strength of 25 MPa at 28 days using 20 mm maximum aggregates size and ordinary Portland cement. Coarse aggregates from crushed stones were subjected to sieve analysis to achieve a ratio of 1:2 for 10 mm and 20 mm respectively for use in all concrete castings.

Slump testing was done on fresh concrete (see Figure 4). 150 mm concrete cubes were prepared, compacted (see Figure 5), de-moulded 24 hours after casting (see Figure 6) and cured in a water tank at 200°C ± 20°C for 7 days, 14 days and 28 days. Compressive strength testing of the concrete cubes was carried out in accordance with ASTM C39-90 [19] (see Figure 7).

RESULTS AND DISCUSSION

Texture and Particle Shape Results

Twenty six sand samples were subjected to texture and shape examination. 52% of the tested sand samples had rough texture compared to 26% that portrayed smooth and fine texture and 22% with rough and fine texture (see Figure 8(a)). 85% of the tested samples were observed to have irregular shaped particles while the rest had rounded shaped particles (see Figure 8(b)). Particles with rough and angular surfaces bind more securely with cement paste and course aggregates compared to the smooth and round shaped particles. Reasonable effect on compressive strength is realized when the slump is widely varied. Angular particles are known to require more water to achieve similar workability compared with the smooth particles.

Irregular and angular sand particles are common in river sands as a result of wave action and attrition forces in water. On the other hand, rounded particles are found in sand pits found on land where sand is mined.

Figure 4. Slump testing.

Figure 5. Compaction of fresh concrete.

Figure 6. De-moulding of concrete cubes.

Figure 7. Compressive strength testing.

Surface texture

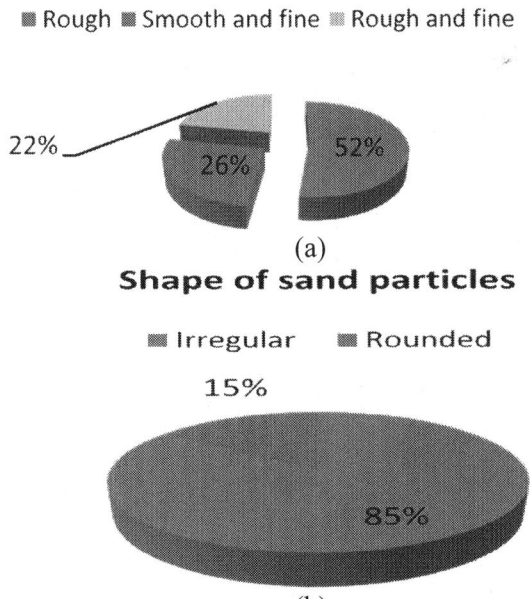

(a)

Shape of sand particles

(b)

Figure 8. Texture and shape of sand particles.

Fineness of Sand

Sand samples were graded using the IS sieves [16] and were categorised into zones as shown in Table 1, Table 2 and Figure 9.

Majority of sand samples (67%) were within Zone II of geological grading implying normal sand. A significant 7% of the tested samples comprised of very fine sand. Such fine grading requires proper mix design proportions to ensure that the quality of resulting concrete is not compromised. Sieve size 600 microns was used to determine degree of fineness in sands and soils.

Results indicate that 26% (7 out of 27) of the samples had over 60% of the samples passing sieve size 600 microns (see Table 2 and Figure 10). This implies that a significant 84% comprised of fine sand samples. Comparatively, 66% of the tested sand samples had over 50% of the samples passing the same sieve.

Silt and Clay Content in Sand

Based on the 27 sand samples tested, the maximum silt and clay content was 42% for NR1 sample compared with the minimum 3.3% for TK1 sand sample (see Figure 11). CL1 (clean sample 1) and CL2 (clean sample 2) were clean control river sand samples that were washed using clean water and sun dried. They had 0.7% and 0.3% silt and clay content after washing. CL2 was used in casting of concrete cubes because it had the lowest level of silt and clay impurities, and organic impurities hence selected to be the control sample.

BS 882 recommends that no more than a maximum of 4% silt and clay content for fines aggregates be used in concrete production.

Grading of sand

■ Zone I: Course sand   ■ Zone II: Normal sand

■ Zone III: Fine sand   ■ Zone IV: Very fine sand

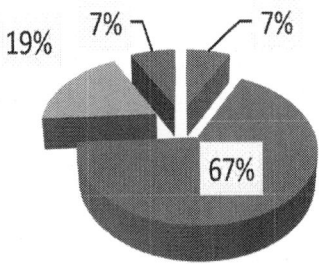

Figure 9. Zoning of sand samples based on fineness.

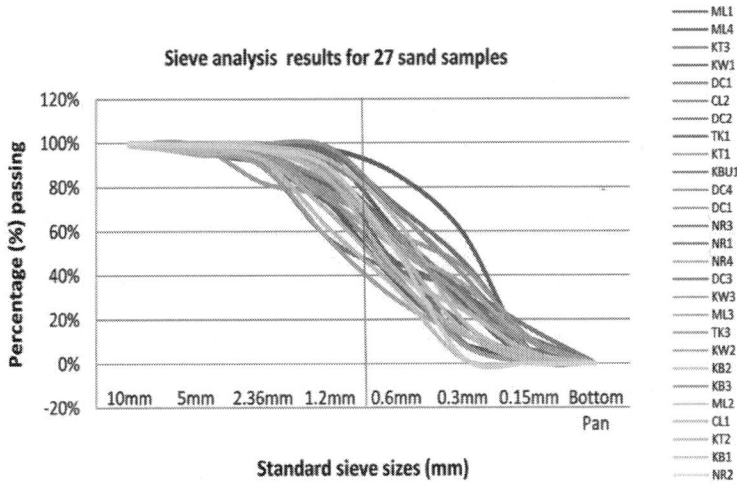

Figure 10. Sieve analysis results.

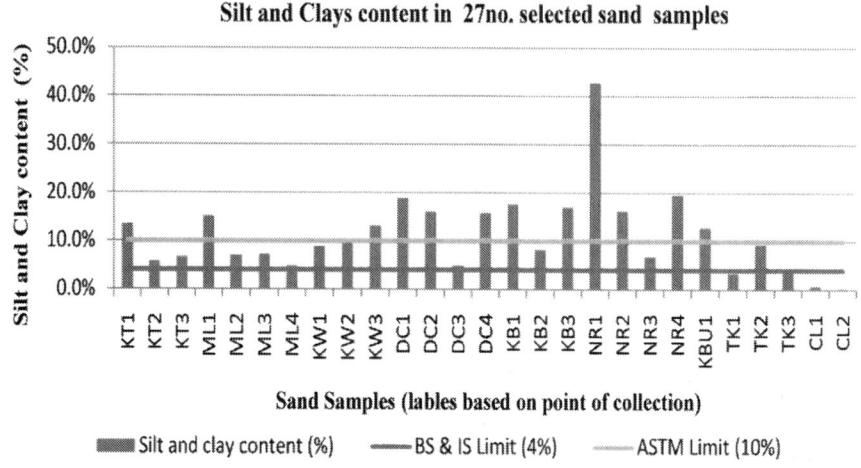

Figure 11. Silt and clay content in sand.

Table 2 Sieve analysis, degree of fineness and grading zone classification for 27 sand samples.

Sample No.	Percentage passing standard sieve sizes (%)													
Sieve size	1	2	3	4	5	6	7	8	9	10	11	12	13	14
Sieve size	ML1	ML4	KT3	KW1	DC1	CL2	DC2	TK1	KT1	KBU1	DC4	DC1	NR3	NR1
10 mm	99	100	100	100	100	100	100	100	100	100	100	100	100	100
5 mm	98	96	96	100	100	100	100	97	99	99	100	100	99	98
2.36 mm	97	92	93	99	100	99	100	94	96	94	98	100	97	89
1.2 mm	93	73	81	97	98	87	95	81	82	72	90	98	88	56
0.6 mm	60	36	51	86	70	57	72	54	53	46	69	70	59	42
0.3 mm	27	9	10	59	40	32	47	32	31	32	40	40	32	34
0.15 mm	2	1	1	5	6	2	11	3	3	11	3	6	3	16
Bottom pan	0	0	0	0	0	0	0	0	0	0	0	0	0	0
Degree of fineness (% passing 600 microns sieve)	60	36	51	86	70	57	72	54	53	46	69	70	59	42
Grading zone classification	III	II	II	IV	III	II	IV	II	II	II	III	III	II	II

Table 3 . Characteristics of samples that failed on compressive strength.

20	NR3	2557.1	1739.96	97.24	0.81	14.14	0.78	0.37	1.35	0.18	387.18	0.11	0.26
21	NB4	2437.70	248.80	49.54	1.05	3.11	6.72	0.89	0.42	0.18	31.98	0.19	0.26
22	KBU 1	250.47	2008.32	158.27	1.20	61.82	5.02	5.00	3.99	0.44	16.25	0.11	8.68
23	TK1	2972.45	1454.78	885.45	1.71	5.19	13.53	0.19	2.04	0.18	17.29	0.11	0.79
24	TK2	2646.51	1148.72	18.04	0.82	2.90	2.53	0.19	0.22	0.18	8.24	0.24	0.26
25	TK3	5709.47	767.78	32.72	1.24	6.40	0.78	0.19	0.87	0.18	11.32	0.11	0.26
26	CL1	5030.75	216.63	13.67	1.46	6.40	5.41	1.17	0.12	0.18	13.54	0.11	0.26

Table 4. Compressive strength for selected samples with constant workability classification.

Sample	Strength at 28 days (MPa)	Silt and clay content (%)	Organic impurities (photometric ohms)	Workability	Source of sand (as per suppliers)	Particle texture and shape
1 ML4	21.552	4.8	0.106	56 mm	Kajiado	Smooth & fine, Irregular
2 KB2	24.403	8.2	0.202	79 mm	Kajiado	Rough & fine, Irregular
3 KT1	24.537	13.5	0.168	93 mm	Machakos	Rough, Irregular
4 NR1	24.937	42.8	0.594	3 mm	Kangundo	Rough, Smooth
5 KW1	24.955	8.8	0.238	4 mm	Machakos	Rough & fine, Irregular

Only four samples out of 27 samples met this limit, representing only 14.8%. An overwhelming 86.2% failed to meet the standard set in BS 882. Comparatively, the ASTM's allowable silt and clay content in sand used for concrete production is 10% by weight. 15 samples met this limit, implying a failure rate of 44.4% of the tested sand samples by ASTM's standard.

From Figure 11, the maximum silt and clay content registered from 27 samples was a significant 42%. This implies that for one tonne of sand, 420 kg is composed of silt and clay impurities. Therefore when such sand is bought for construction, value for money is not achieved since over half of the sand quantity comprises of silt and clay impurities.

Organic Impurities in Sand

With regard to testing for organic impurities in sand, the standard requires that the color of sodium hydroxide solution in sand should be lighter than the solution of sodium hydroxide mixed with tannic acid, both solutions having been preserved for 24 hours after mixing as detailed in ASTM C40 and IS standard [16] . Out of the 27 sand samples tested, only 6 samples indicated lighter color than the standard solution 24 hours after mixing. This indicates that 23% of the collected were within the organic content limit as set in ASTM C40, indicating a failure rate of 77%.

Further, color analysis was carried out using photometric equipment and results are shown in Figure 12. It was found that the maximum value of photometric resistance for clean sample was 0.205 ohms for CL1. CL2 recorded the lowest color resistance of 0.023 ohms indicating the lowest level of organic impurities. Consequently assuming 0.205 ohms for the washed sand sample to be the upper limit for organic impurities, only 13 samples indicated value of less than 0.205. This implies over 50% of the sand samples exceeded the maximum organic content for the washed control sample.

Combination of Silt and Clay Content and Organic Impurities for Selection of Test Samples

Based on the results for levels of impurities obtained for the 27 sand samples, 13 samples with varying level of impurities were selected for casting of concrete cubes for compressive strength testing. To ensure even distribution of sand samples of various levels of impurities in the final set of samples selected for casting, samples were categorized under classes of pre-set ranges of levels of impurities starting with the lowest intervals of 5%. Results of silt and clay content 1% - 5%, 5% - 10%, 10% - 15%, 15% - 20% and 20% - 50% while organic impurities were categorized into classes of 0.2 - 0.3, 0.3 - 0.4, 0.4 - 0.5, 0.5 - 0.6, 0.6 - 0.7, and 0.7 - 0.8 ohms. In the selection process for the final list of samples, a minimum of 30% of the samples falling in each class was chosen to ensure fair representation from each class for silt and clay content as well as organic impurities levels. Where 30% was not achieved, the process entailed replacement of the sand sample until this representation was achieved. Since the results obtained from 27 samples were within the range of above classes and due limitation of standard concrete casting molds, cost and time, 13 samples selected for casting of concrete are shown in Figure 13.

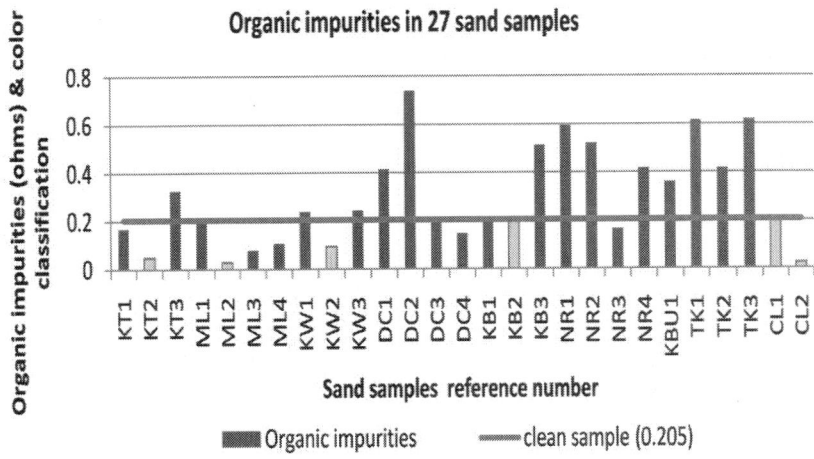

Figure 12. Organic impurities in 27 sand samples.

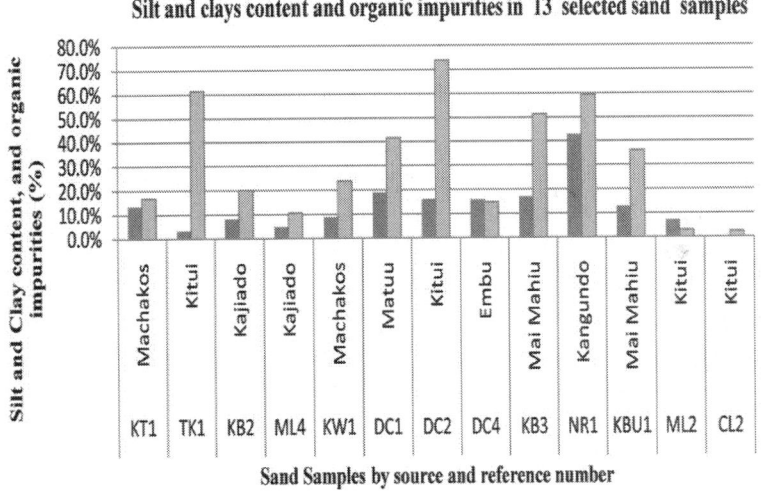

Figure 13. Organic impurities and silt and clay content in 13 selected sand samples.

The 13 sand samples selected for casting of concrete cubes were a good representative of the 27 samples collected.

Specific Gravity of Sand Samples

Sand samples were subjected to specific gravity tests as detailed in IS standard [16] equivalent to ASTM D854 [19] for aggregates less than 10mm diameter using the pycnometer glass vessel. Results showed that the average apparent specific gravity was 2.7 while the average water absorption of dry mass was 2.9. This compares well with the expected specific gravity values of 2.7 for sand used in concrete production, implying that the sand used in this research represent the commonly used normal sand used in concrete making. This indicates that the sand samples used were within the normal range for building sand. Bulk specific gravity is used for calculation of the volume occupied by the aggregate in various mixtures such as concrete. Apparent specific gravity pertains to the relative density of the sand making up the constituent particles not including the pore space within the particles that is accessible to water. Bulk density varied from 2.54 to 2.81 for TK2 and KT3 respectively. This explain why the slump observed and water absorption by pores was specific to a particular sand sample based on mode of sample formation e.g. river sand and pit sand.

Compressive Strength of Concrete for Various Levels of Silt, Clay and Organic Impurities

13 samples were selected for cube compressive strength testing according to ASTM C39-90 [20] and BS 1881 [18] . For each sample, a total of 9 cubes were cast and cured under water at room temperature. Three concrete cubes made from each sand sample were tested at the age of 7 days, 14 days and 28 days after casting using a universal testing machine. The average was obtained from 3 cubes tested and results are as shown in Figure 14. The expected compressive strength at day 7 (E7DS), day 14 (E14DS) and day 28 (E28DS) are also shown.

It is important to note that a uniform mix design used for most low rise structural buildings was adopted, that is 1:1.5:3:0.57 for cement: fine aggregates: coarse aggregates: water. For KBU1 sample, the slump was zero (too stiff, extremely low) hence water: cement ratio adjusted to 0.58 hence slump of 9mm was obtained. KBU1 was made from volcanic pit sands and it was observed that it requires more water during mixing to achieve medium to low workability levels. It was noted that this sample requires more water to achieve normal and had irregular shaped and rough texture.

From the above results, three samples (that is NR1, ML4 and KB2) failed to meet the minimum strength expected at day 7, one sample (ML4) failed at day 14 and 5 samples (NR1, KW1, KT1, ML4, KB2) failed to meet the compressive strength expected at day 28. This represents 38% failure rate at 28 days. Since all the samples were subjected to similar casting and curing conditions, this failure is largely attributed to the presence of silt and clay content and organic impurities in sand and to some extent to particles shapes, sizes and texture. It was observed that all the samples that failed at day 7 and day 14 also failed at day 28. However not all samples that failed at day 28 had indicated failure at day 7 and day 14. These include KW1 and KT1 that had passed the strength requirement at day 7 and 14 but failed at day 28. This affirms the importance of concrete strength testing up to 28 days maturity.

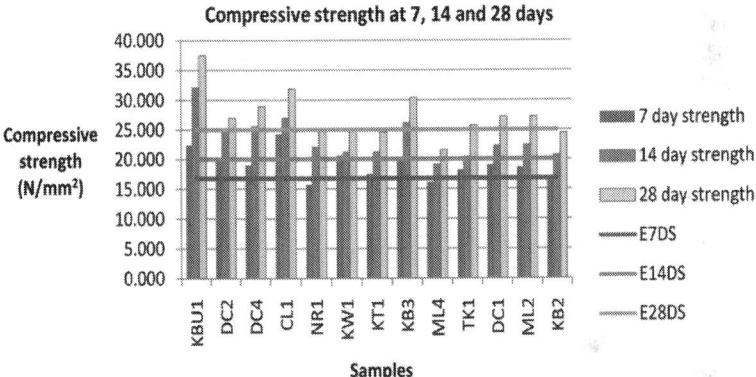

Figure 14. Compressive strength of concrete cubes at age of 7, 14 and 28 days.

Table 3 shows the characteristics of samples that failed. It is deduced that the lowest level of impurities is 4.8% while the lowest level for organic impurities was 0.106 ohms photometric color classification. It can be taken that any sample having of 4.8% silt and clay content and 0.106 ohms for organic impurities or more is likely to fail. Two (ML4 and NR1) of the five samples that failed on compressive strength testing had smooth particles implying that particles sizes play some role in concrete compressive strength. It is noted that 3 (KB2, KT1 and KW1) of the failed samples that portrayed rough and irregular particles had higher silt and clay content of more than 8%.

Smooth and round sand surfaces provide a weak interlocking bond between cement and course aggregates thus contributing to reduced compressive strength of concrete. Since 3 out of 5 samples that failed to meet the expected compressive strength at day 28 had rough texture and irregular shape particles, it implies that besides silt and clay and organic impurities in sand, particle sizes and shapes form a significant factor in determination of compressive strength of concrete.

A Case of Compressive Strength with Constant Workability

In order the assess the effect of workability on the compressive strength of concrete made from selected sand samples having varying level of silt and clay and inorganic impurities, 4 sand samples were cast while maintaining workability constant. A set of concrete cubes was cast while maintaining workability to be within Very Low (0 - 25 mm) category as shown by KBU1 (a), DC 2 (a), CL2 (a) and DC4 (a). A second set was made from the same sand samples was cast using a constant workability of Medium (50 - 100 mm) category as shown by KBU1 (b), DC 2 (b), CL2 (b) and DC4 (b). The results are presented in Table 4 and Figure 15.

Table 5. Categorization of compressive strength data for regression analysis.

Sample	Silt and clay content (%)	Organic Impurities (ohms)	7 day strength (N/mm²)	14 day strength (N/mm²)	28 day strength (N/mm²)	Slump (using cone)	Workability classification	Source
1a KBU1	12.8	0.361	22.45	32.13	37.45	9 mm	Very low	Mai Mahiu
1b KBU1	12.8	0.361	12.93	16.55	20.73	86 mm	Medium	Mai Mahiu
2a DC2	16.1	0.738	19.98	25.05	26.98	11 mm	Very low	Kitui
2b DC2	16.1	0.738	16.19	19.35	24.05	69 mm	Medium	Kitui
3a CL2	0.3	0.023	24.23	26.95	31.86	13.5 mm	Very low	Kitui
3b CL2	0.3	0.023	16.05	19.20	24.68	58 mm	Medium	Kitui
4a DC4	15.8	0.147	19.02	25.52	28.93	10 mm	Very low	Embu
4b DC4	15.8	0.147	12.40	16.20	19.62	65 mm	Medium	Embu

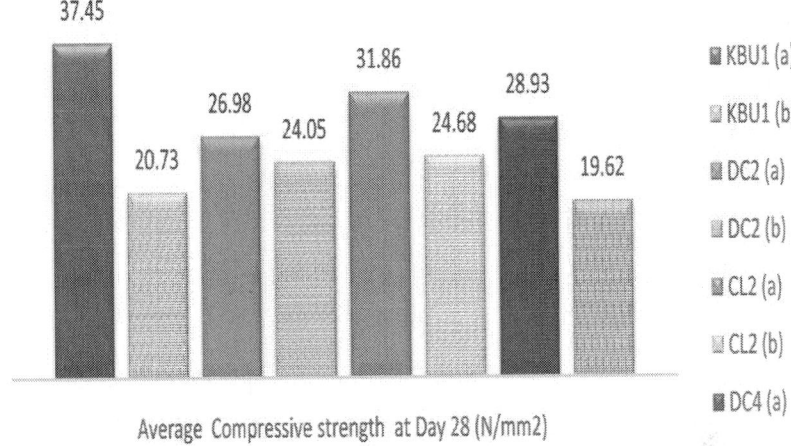

Figure 15. Compressive strength for samples with constant workability.

Table 6. Regression analysis sand impurities and compressive strength.

Sample	28 day strength (N/mm²)	Silt and clay content (%) (x1)	Organic impurities (ohms) (x2)	Sand texture	Particles shape	Grading	Slump
(a) Group of sand samples with rough surface texture, moderate grading & high workability							
1 KB2	24.403	0.082	0.202	Rough	Irregular	65%	79 mm
2 KT1	24.537	0.135	0.168	Rough	Irregular	53%	93 mm
3 KBU1	20.730	0.128	0.361	Rough	Irregular	46%	86 mm
4 CL2	24.680	0.003	0.023	Rough	Irregular	57%	58 mm
5 DC4	19.620	0.158	0.147	Rough	Irregular	69%	65 mm
(b) Group of sand samples with rough surface texture, moderate grading & low workability							
6 KW1	24.955	0.088	0.238	Rough	Irregular	86%	4 mm
7 DC4	28.929	0.158	0.147	Rough	Irregular	69%	10 mm
8 CL2	31.858	0.003	0.023	Rough	Irregular	57%	13.5 mm
9 KBU1	37.455	0.128	0.361	Rough	Irregular	46%	9 mm
(c) Group of sand samples with smooth surface texture, varied grading & varied workability							
10DC2	26.979	0.161	0.738	Smooth	Rounded	72%	11 mm
11KB3	30.393	0.17	0.513	Smooth	Rounded	58%	5 mm
12DC1	27.188	0.189	0.415	Smooth	Rounded	70%	52 mm
13NR1	24.937	0.428	0.594	Rough	Rounded	42%	3 mm

It is clear from the below results that workability plays significant role in determination of compressive strength of concrete. By varying slump from very low (0 - 25 mm) to medium (50 - 100 mm), it was observed that compressive strength reduced with a margin of between 17 N/mm^2 for KBU1 to 2 N/mm^2 for DC2.

At very low slump, KBU1 depicted the highest compressive strength of 37.45 N/mm^2 compared with DC2 that registered the lowest compressive strength of 20.73 N/mm^2. On the other hand, at medium slump CL2 registered the highest compressive strength while DC4 registered the lowest strength. Comparatively DC2 had the minimal strength effects of 2.9 N/mm^2 with changes in the slump level while KBU1 had the largest effect of 16.7 N/mm^2 on compressive strength with changes in slump level. This implies that DC2 is a better sand sample since significant savings can be made from reduction in the quantity of water used during casting without significant changes in compressive strength.

KBU1 had rough and irregular shaped particles while DC2 had smooth and sand rounded particles. DC4 had rough and fine particles that were irregular in shape while CL2 had irregular shaped rough particles. This implies that effect of workability on the compressive strength of concrete is more pronounced in rough and irregular shaped sand particles in comparison with rounded and smooth sand particles.

Analysis of the Correlation between Compressive Strength and Sand Impurities

In order to assess the relationship between compressive strength obtained from concrete made using sand with different levels of silt and clay content and organic impurities the compressive strength results were categories into 3 as shown in the Table5 A constant water cement ratio of 0.57 was used. The samples were categorized based on similar or closely related characteristic of surface texture, particle shapes, slump level and degree of fineness obtained from percentage passing 600 microns sieve size.

The samples used for regression analysis were carefully selected to ensure that their properties are almost similar or as close as possible and that only silt and clay and organic impurities significantly varied all other factors being held constant. By use of 5 sand samples which had similar texture, particles shapes and closely linked grading curves as indicated is group (a) in the table above regression analysis was used to derive the relationship between compressive strength of concrete cubes with varying silt and clay content as illustrated in Figure 16.

A regression equation for predicting compressive strength of concrete made from sand containing varying level of silt and clay contents was found to be:

$$y = ax_1 + bx_2 + c \tag{1}$$

$$Fcu28 = -23.20SCI - 2.416ORG + 25.57 \tag{2}$$

with $R^2 = 0.444$

where Fcu28 = cube compressive strength at day 28;

SCI = silt and clay content in sand;

ORG = Organic impurities content in sand.

The output from regression and correlation analysis showing the relationship between silt and clay content and organic impurities against compressive strength of concrete are shown in Table 6 and Table 7.

From the R^2 value, it is deduced that the contribution of the silt and clay content and organic impurities to the overall compressive strength of concrete is a significant 44%. This implies that although there are other factors contributing to the compressive strength of concrete, presence of silt and clay content and organic impurities plays a major role.

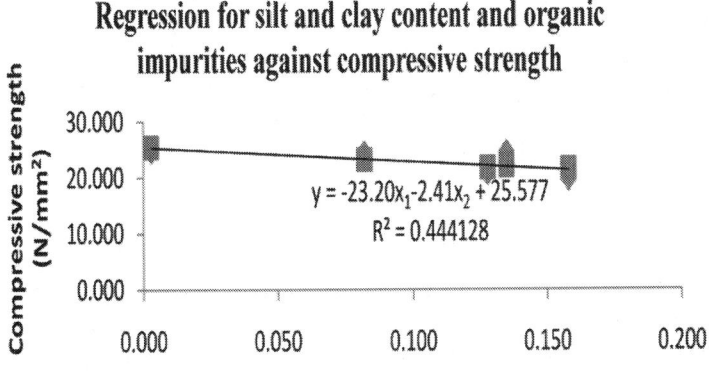

Figure 16. Regression analysis for compressive strength relationship.

Note: Grading is measured in terms of percentage passing 600microns standard sieve.

Table 7 Correlation analysis between silt and clay content and organic impurities and compressive strength.

Multiple R	0.666429		Coefficients	Standard error	t-statistic	P-value
R²	0.444128	Intercept	25.577	2.530	10.111	0.010
Adjusted R²	−0.11174	SCI (Variable 1)	−23.203	25.720	−0.902	0.462
Standard Error	2.556553	ORG (Variable 2)	−2.417	12.997	−0.186	0.870

	y	x₁ (SCI)	x₂ (ORG)
y	1		
x₁ (SCI)	−0.65918	1	
x₂ (ORG)	−0.46683	0.587901	1

The other factors may include mode of sand formation and workability, workmanship, quality of course aggregates and quality of water among others. Therefore concrete designers must provide adequate factor of safety to guard against structural failure as a result these impurities in building sand. Frequent testing of sand for construction purposes is therefore highly recommended to ensure measures are put in place e.g. washing of sand in a bid to prevent collapse of buildings as a result of excessive levels of silt and clay content and organic impurities.

From Table 7, it is deduced that contribution of silt and clay content (SCI) toward the compressive strength of concrete is a significant 65%. Similarly the contribution of organic impurities (ORG) toward compressive strength of concrete is 46%.

It is observed that the contribution of silt and clay content towards compressive strength of concrete is more significant compared with the organic impurities.

From the figure above, it is clear that increase in silt and clay content and organic impurities significantly reduces the compressive strength of concrete. The 44% contribution represents a contribution of $11N/mm^2$ for concrete with target strength of $25 N/mm^2$. Therefore present of these impurities cannot be ignored during concrete production process and they may lead to failure and collapse of structural buildings. Besides presence of silt and clay content and organic impurities having a significant contribution to buildings failure, other factors such as specific gravity, curing and workability, workmanship, adherence to structure designs, works supervision and quality of other concrete ingredients plays important role in buildings failure.

CONCLUSIONS AND RECOMMENDATIONS

From the study, it is observed that building sand being supplied in Nairobi City County and its environs contained silt and clay contents, and organic impurities that exceeded the allowable limits. The level of silt and clay content ranged from 42% to 3.3% for while organic impurities ranged from 0.029 to 0.738 photometric ohms unwashed sand. An overwhelming 86.2% of the tested sand samples failed to meet silt and clay content limits set out in BS 882 while 44.4% exceeded the limit set out in ASTM limits. With regard to organic content, 77% of the sand samples studied exceeded the recommended organic content for concrete production by ASTM standard. A total of 38% of the concrete cubes made from sand with varying sand impurities failed to meet the design strength of 25 Mpa at age of 28 days. It observed that the allowable minimum level of silt and clay content and organic impurities in sand being supplied in Nairobi and its environs is 4.8% and 0.106 ohms respectively. Beyond these limits then the resultant concrete will fail to meet the expected strength at 28 days age. It is this concluded that the presence of impurities in sand significantly contributed to reduction in compressive strength of concrete strengths which may lead to collapse of buildings if not addressed in the concrete design mix.

Regression equation $Fcu28 = -23.20SCI - 2.416ORG + 25.57$ with $R^2 = 0.444$ was generated to predict the compressive strength of concrete with varying levels of silt and clay contents, and organic impurities respectively. It is noted that 44% of compressive strength is contributed by silt and clay content and organic impurities in sand used for concrete production. Since presence of these impurities significantly affect the compressive strength of concrete, they cannot be ignored hence the need to ensure sand free from these impurities is used during concrete production. This equation is applicable to concrete made using building sand with similar physical and chemical properties as the samples tested.

It was observed that 3, 1 and 5 samples failed on compressive strength at 7, 14 and 28 days. This study recommends monitoring of strength at 56 days and beyond to establish any trend beyond 28 day. Sand samples were sourced from supply points in Nairobi County and its environs. It is appreciated that some suppliers do adulterate sand by mixing it with soils for unjustified economic gains. This was evident during samples collection where it was observed that some suppliers received sand from different sources and mixed it up to dilute the negative color, texture and silt and clay content levels. This study recommends further study to establish the quality of sand collected directly from the source (river, pit or sea) in comparison to the quality of sand sourced at the supply points (market places) in order to establish the extent of adulteration within the supply chain.

In regard to construction management practices, construction professional are to enhance inspection of the quality of building materials to ensure that quality, cost, time and customer expectations for concrete structures is not compromised and to avoid the collapse of buildings as observed in Nairobi in the recent years. It is noted that investors lose up to 40% of their investment through purchase of sand with impurities. Kenya and other developing countries need to formulate policies to govern allowable limits of silt and clay and organic impurities in sand and ensure that materials are inspected and approved by an authorized construction professional before use.

ACKNOWLEDGEMENTS

We would like to thank Pan African University Institute for Basic Sciences, Technology and Innovations for the scholarship and funding of this research work.

REFERENCES

1. Orchard, D.F. (1979) Concrete Technology, Propertiess of Material. 4th Edition, Volume 1. Applied Science Publishers Ltd., London, 139-150.
2. Machuki, O.V. (2012) Causes of Collapse of Buildings in Mombasa County. A Case of Mombasa City—Kenya. Published on Department of Extra Mural Studies. University of Nairobi, Nairobi, Kenya. http://ems.uonbi.ac.ke
3. Ayodeji, O. (2011) An Examination of the Causes and Effects of Building Collapse in Nigeria. Journal of Design and Built Environment, 9, 37-47. http://e-journal.um.edu.my/filebank/published_article/3294/Vol%209-3.pdf

4. Ayuba, P., Olagunju, R. and Akande, O. (2011) Failure and Collapse of Buildings in Nigeria: Roles of Professionals and Other Participants in the Building Industry. Interdisciplinary Journal of Contemporary Research in Business, 4, 1267-1272.

5. Dimuna, K.O. (2010) Incessant Incidents of Building Collapse in Nigeria: A Challenge to Stakeholders. Global Journal of Researches in Engineering, 10, 75-84.

6. Dahiru, D., Salau, S. and Usman, J. (2014) A Study of Underpinning Methods Used in the Construction Industry in Nigeria. The International Journal of Engineering and Science (IJES), 3, 05-13. http://www.theijes.com/papers/v3-i2/Version-3/B03203005013.pdf

7. Oloyede, S., Omoogun, C. and Akinjare, O. (2010) Tackling Causes of Frequent Building Collapse in Nigeria. Journal of Sustainable Development, 3, 127-132.http://dx.doi.org/10.5539/jsd.v3n3p127

8. Savitha, A. (2012) Importance of Quality Assurance of Materials for Construction Work. Building Materials Research and Testing Division, 1-5.

9. Olanitori, L.M. (2006) Mitigating the Effect of Clay Content of Sand on Concrete Strength. 31st Conference on Our World in Concrete & Structures, Singapore, 16-17 August 2006.

10. Olanitori, L.M. and Olotuah, A.O. (2005) The Effect of Clayey Impurities in Sand on the Crushing Strength of Concrete (A Case Study of Sand in Akure Metropolis, Ondo State, Nigeria). 30th Conference on Our World in Concrete and Structures, Singapore, 23-24 August 2005.

11. BS 882 (1992) Specification for Aggregates from Naturanl Sources for Concrete. British Standard.

12. ASTM C117 (1995) Standard Test Method for Materials Finer than 75-um (No.200) Seive in Mineral Aggregates by Washing. American Society for Testing Materials, West Conshohocken.

13. Construction Standard CS3 (2013) Aggregates for Concrete, Technology, Ed., The Government of the Hong Kong Special Administrative Region, Hong Kong.

14. Harrison, D.J. and Bloodworth, A.J. (1994) Construction Materials, Industrial Minerals Laboratory Manual. Technical Report WG/94/12, Nottingham.

15. Anosike, N.M. (2011) Parameters for Good Site Concrete Production Managment Practice in Nigeria. Ph.D. Thesis, Deparment of Building Technology, College of Science & Technology, Covenant University, Ota.

16. IS 383 (1970) Specification of Course and Fine Aggregates from Natural Sources for Concrete. Bureau of Indian Standards, India.

17. ASTM C40 (2004) Standard Test Method for Organic Impurities in Fine Aggregates for Concrete. ASTM International, West Conshohocken.

18. BS 1881-120 (1983) Testing Concrete Method for Determination of Compressice Strength of Concrete Cores. British Standards Institute, London.
19. ASTM D854 (2014) Standard Test Method for Specific Gravity of Soil Solids by Water Pycnometer. ASTM International, West Conshohocken.
20. ASTM C39 (1990) Standard Method of Test for Compressive Strength of Concrete Specimens. ASTM International, West Conshohocken.

CITATION

Ngugi, H. , Mutuku, R. and Gariy, Z. (2014) Effects of Sand Quality on Compressive Strength of Concrete: A Case of Nairobi County and Its Environs, Kenya. Open Journal of Civil Engineering, 4, 255-273. doi: 10.4236/ojce. 2014.43022.

Index